Herbartism in Austrian Philosophy

Meinong Studies/
Meinong Studien

Edited for

Alexius-Meinong-Institut – Forschungsstelle und Dokumentationszentrum für Österreichische Philosophie, Karl-Franzens-Universität Graz

by Mauro Antonelli, Marian David

Volume 11

Herbartism in Austrian Philosophy

Edited by
Carole Maigné

DE GRUYTER

ISBN 978-3-11-074729-4
e-ISBN (PDF) 978-3-11-074732-4
e-ISBN (EPUB) 978-3-11-074744-7
ISSN 2198-2309

Library of Congress Control Number: 2021940041

Bibliographic information published by the Deutsche Nationalbibliothek
The Deutsche Nationalbibliothek lists this publication in the Deutsche Nationalbibliografie; detailed bibliographic data are available on the Internet at http://dnb.dnb.de.

Printing and Binding: LSC Communications, United States

www.degruyter.com

Contents

Carole Maigné

Introduction: Austrian Herbartism, Herbartism in Austrian Philosophy

One of the difficulties of working on Herbartism in Austria-Hungary is the fight against the obligatory reference: quoting Herbart and his school is frequent, reading them attentively is less frequent. So, the researcher is somehow confronted with a "well-known" reference, in the sense given to it by one of his sworn enemies, Hegel: "What is familiar and well known as such is not really known for the very reason that it is *familiar and well known*."[1] However, as the history of the Austrian tradition and theoretical reflections in this field expand, discussion of the texts read is becoming more and more tight and precise. And when the reference emerges and is discussed, it must be contextualized, depending on what is defined as "Austrian". Because Herbartism reached its peak in the second half of the nineteenth century, and was effectively institutionalized as "official philosophy" of the Austro-Hungarian Empire,[2] at least in Prague and Vienna, criticizing Herbartism often means discussing the "Austrian", "philosophical" and "institutional" criteria of the object under consideration.[3]

Herbart's realism[4] is original and at the heart of his success in the Austrian tradition. He credits Kant with asking the key question of philosophy: "*wie entstehen für uns Gegenstände?*"[5] But where Kant sought the elaboration of our

1 "Das Bekannte überhaupt *ist* darum, weil es *bekannt* ist, nicht *erkannt*." (Hegel 1807, p. 35.)

2 Herbart's disciples in the Austro-Hungarian Empire: Hermann Bonitz, Franz Karl Lott, R. Zimmermann, Theodor Vogt, J. W. Nahlowski, F. Cupr, W. F. Volkmann, Ritter von Volkmar, G. A. Lindner, A. Drbal, J. Dastich, J. Durdík, O. Hostinský (Koschnitzke 1988).

3 Working on Herbartism also leads to a re-evaluation of the history of German philosophy: Beiser 2014.

4 Herbart 1806 defines his philosophy as "realistic metaphysics" (*realistische Metaphysik;* § 9) or "strict realism" (*strenger Realismus*; § 14).

5 Herbart 1828, p. 56: "Die grosse Frage: *wie entstehen für uns Gegenstände?* war nun erhoben; die alte Voraussetzung, die Dinge seien da, und liessen sich durch die ontologischen Prädicate erkennen, war für einen consequenten Denker auf immer in ihrer Ruhe gestört. Früher nahm man Begriffe sowohl als Dinge, wie man sie eben fand; jetzt waren die einen und die andern für uns, in uns, durch uns."

Carole Maigné, University of Lausanne
carole.maigne@unil.ch

https://doi.org/10.1515/9783110747324-001

knowledge (*Bearbeitung der Erkenntnisse*),[6] Herbart seeks the elaboration of concepts (*Bearbeitung der Begriffe*), a definition he gives of philosophy.[7] Where Kant puts a legislative reason in front of nature, Herbart denounces the psychology of faculties (*Vermögenspsychologie*) and promotes a critique of the concepts of experience. For Herbart, the work of philosophy is thus twofold: on the one hand, to propose an ontology that allows the grasping of the real up to its ultimate elements (*Realen*), and on the other hand, to propose an explanation of the emergence of our concepts, namely a psychology to counter the transcendental Kantian subject. The idea of a conceptual construction is for him quite crucial,[8] the philosopher even develops an "art of construction" (*Kunst der Construction*).[9] The analytical reconstruction of common sense draws all the consequences of the Kantian Copernican revolution, where ontology becomes an analytic of understanding, but without reducing the ontological question to a linguistic problem.[10] Cassirer rightly considered Herbart to be "a master in the discovering of the dialectic of perceiving consciousness"[11]. Herbart would like to find a way to structure the given of experience from within, even though it is no longer possible to dream of the direct adequacy between thing and idea. He thinks in the wake of Kant but defines oneself as a "Kantian of 1828" (a Kantian working after the work of Kant and after the so-called "fall" of German idealism).[12] Constructing experience implies inventing a "method of relations", an "art of contingent perspectives".[13] The model is mathematical: the object is defined as a function, that is, an order, a well-ordered series that places its perfect completeness at the limit, what we have often called, following G. Preti and R. Pettoello, a "new sense of the transcendence of the object" in philosophy, or "objectless realism" to use R. Pettoello's expression.[14] The thing itself remains unknowable to us, but there is nothing to be

6 Kant, *Kritik der reinen Vernunft*, B VII, p. 40.

7 Herbart 1813–1837, § 1.

8 Herbart 1829, § 163, p. 12: "Jede Speculation, sie heisse nun Theorie, System, oder wie man will, sucht eine Construction von Begriffen, welche, wenn sie vollständig wäre, das Reale darstellen würde, wie es dem, was geschieht und erscheint, zum Grunde liegt."

9 Herbart 1829, § 191.

10 Pettoello 2002, p. 38.

11 Cassirer 2000, p. 371.

12 Herbart, 1828, Vorrede, p. 13: [Against Spinoza] "lehrt Kant: 'unser Begriff von einem Gegenstande mag enthalten, was und wieviel er wolle: so müssen wir doch aus ihm herausgehn, um diesem die Existenz beyzulegen.' Dieses nun ist der Hauptpunct, auf welchen das vorliegende Buch überall hinweiset; und darum ist der Verfasser Kantianer, wenn auch nur vom Jahre 1828, und nicht aus den Zeiten der Kategorien und der Kritik der Urtheilskraft."

13 Herbart 1829, § 190, p. 45.

14 Pettoello 2001, 2002, 2003; Maigné 2005.

regained, there is no intrinsic lack of knowledge. It is rather a question of reformulating the very concept of transcendence by considering that the logical-mathematical relations are the place of the object, are its immanence. Herbartian realism thus defines metaphysics as the "science of the comprehensibility of experience (*Begreiflichkeit der Erfahrung*)"[15]. We also understand the importance of pedagogy as education and orientation of this experience of the giveness: for Herbart it is applied metaphysics. We also guess the importance of aesthetics, in the sense of *aesthesis*: there is no romanticism of beauty in Herbart's work, no revelation of the absolute in the work of art, but a questioning of relations between elements, of the way relations are linked together. Herbart's Philosophy is a philosophy of the *Wie?*

What Hoeschen and Schneider refer to as "Herbarts Kultursystem", and which we would gladly take up in this Austrian and not only German context, is the spread of Herbartism in all the disciplines of the time (from mathematics to psychology, art, literature and linguistics): There is a systematicity of knowledge in Herbartism which is not that of a system closed in on itself, the unity comes from the functional relationships between the different fields of knowledge, a unity therefore where the object is "constructed", according to a plurality of points of view.[16] But it is also, and this seems important to us in order to understand the success of Herbartism in the Austro-Hungarian Empire, a reflection on culture. Herbart already spoke of a "system of culture" (*Cultursystem*): culture is for him a living configuration of representations, taken up by the subject but also overflowing it, in a fundamental social and plural intersubjectivity.[17] According to Hoeschen and Schneider, this would paradoxically be one of the reasons for the decline, disavowal and weariness of Herbartism at the end of the 19th century, when it became a common good, in a way too "well known".

One of the crucial vectors of Austrian Herbartism is its aesthetic formalism, carried by the Herbartian school and by Robert Zimmermann in particular, with a remarkable longevity and tenacity over the decades 1850–1890.[18] The internal

---

15 Herbart 1828, § 81.

16 Hoeschen/Schneider 2001 and Maigné 2007. Herbart, in his *Lehrbuch zur Einleitung in der Philosophie* (§ 1) considers moreover that philosophy has no object of its own: "Philosophie besitzt nicht, gleich andern Wissenschaften, einen besonderen Gegenstand, mit dem sie sich ausschließend beschäftigt. Ihre Eigentümlichkeit muss also in der Art und Weise gesucht werden, wie sie jeden sich darbietenden Gegenstand behandelt." Cf. Herbart 1837.

17 Orth 2001 p. 25–37. See also Zumr 1998.

18 Henckmann pointed out that the first occurrence of the "Allgemeine Kunstwissenschaft" is by Zimmermann, as early as 1862, in this crucial text: *Zur Reform der Aesthetik als exacter Wissen-*

and external debates around Herbartian formalism are both constant and virulent, and as such constitute a tradition,[19] opening towards other formalisms of the 20th century, notably Czech and Russian.[20] Echoing the complexity of the links between the protagonists, and in order to give reason both for the affiliations and for the differences in positions, it is not surprising that the format of the anthology of texts is pertinent here.[21] Conscious of itself, this Herbartian tradition was historicized at a very early stage, as is shown by Zimmermann's text, with this fundamental title: *Geschichte der Aesthetik* in 1858, the first of its kind, which in its last chapter, entitled "The Aesthetics of Realism" (*Aesthetik des Realismus*), establishes Herbartian realism and formalism as the future of the discipline and the discipline's self-reflection on itself. In an astonishing combination of Herbart, Bolzano and Hanslick, Zimmermann invented a coherence between all of them such that it is both the culmination of the historical process that he reconstructed and the conceptual ferment of Herbartian formalism, which Zimmermann deployed in *Zur Reform der Aesthetik als Formwissenschaft* (1862) and *Allgemeine Aesthetik* (1865), among others, just after becoming a professor in 1861. This final chapter of 1858 is worthy of the first outline of an Austrian self-reflexive path after 1848. This aesthetic refuses the Kantian judgment of taste, refuses the philosophical and dialectical historicisation of Hegelian art, refuses the recourse to the ineffable, to genius, to intuition, seeking to become a science, against the idealistic and romantic *Schwärmerei*, relying on an anti-psychologism in the aesthetic judgment that rejects sentiment for a reflection on value. In this Austro-Herbartian filiation, Robert Zimmermann is emblematic, but surprisingly little read for himself.[22]

The heart of the diffusion of Herbartism is therefore an aesthetic formalism that spreads to other fields, because aesthetics involves much more than the question of art: it involves an epistemology, an ethics and, as we underlined above, a philosophy of culture. In this respect, let us read what Julius von Schlosser (1866–1938) said in 1934, in his article "Die Wiener Schule der Kunstgeschichte",[23]

---

schaft (Henckmann 1985, p. 277f). See also Rampley 2013, even if Herbartism is not central in his research.

19 Maigné 2012, p. 10–19.

20 Maigné/Trautmann-Waller 2009.

21 Maigné 2012 with translations of Bolzano, Herbart, Zimmermann, Durdik, Hostinský, Zich, Utitz; see also the website http://www.formesth.com/index.php (visited on June 20, 2020), and more recently, but without focus on Austrian Philosophy: Stöckman 2019.

22 The first monograph on Zimmermann is Maigné 2017. K. Blaukopf had considered writing one without having been able to complete this project.

23 Von Schlosser 1934, p. 141–228.

because he in fact draws up a critical panorama of the influence of Herbartism,[24] but also a critique of its expansion. Schlosser considers that

> our Vienna remained the last stronghold of Herbartism until the last third of the old century; something that is not without importance for the history of the Viennese art historical school, since its influence – not to mention the widespread booklet by the music historian Hanslick – can still be felt even in the thinking of A. Riegl.[25]

However, it is worthwhile not to stop there and to read the whole text carefully. Schlosser's anti-portrait of Zimmermann and Herbartism is presented in this historical review, and the arguments deployed are typical of the critics of Herbartism at the beginning of the 20th century who described it as an outdated philosophy. For Schlosser, the Viennese school is a typical example of a successful chapter of German, Hegelian and Prussian science in Austria:[26] it is because Herbartism is obsolete that Austrian renewal is here germanized, as a backlash to the installation of "Austrian" Herbartism against "German philosophy". Paradoxically, the "fortress" has another metaphor to counteract it: Herbartism is a "fluid" (*Fluidum*). Schlosser draws an astonishing portrait of Zimmermann, both everywhere and without imprinting on anyone, he both recognizes and defuses the weight of Austrian Herbartism. The case of Riegl is symptomatic:

> [...] the fluid of Herbartian Realism, which had a long-lasting effect on Austrian school psychology (in reality, by the way, the last great building of classical philosophy in Germany), also overflowed Riegl, and in his later development there might be more than one point in which the aftereffect of Herbart's aesthetic formalism, as represented by the aesthetician Zimmermann in his most recent work, is noticeable.[27]

If Riegl became an art historian, and a Viennese glory, it is thanks to the boredom he felt with Zimmermann's philosophy courses... The refusal of singularity, the promotion of anonymity,[28] the idea that the art historian should even be "without personal taste" testify to the Herbartian abstraction that persists. Schlosser continues: if the *Kunstwollen* is an "objectification of the value judgement", then

---

24 Note that the only article in French on Zimmermann is by another art historian, Lionello Venturi (1885–1961): "R. Zimmermann et les origines de la science de l'art", a contribution delivered at the 1937 International Congress of Aesthetics (Paris). The link with the institutionalization of the discipline is therefore recurrent.
25 Von Schlosser 1934, p. 149.
26 Von Schlosser 1934, p. 145.
27 Von Schlosser 1934, p. 182.
28 Von Schlosser 1934, p. 191.

perhaps it is precisely an extension of the value judgement that Herbartism puts in place in the aesthetic field:[29]

> For here again, from a very high point of view, an attempt has been made to determine the values that have emerged over the centuries within the field of monument conservation in terms of the philosophy of history. And here it must be noted once again that here, as with Riegl's work in general, Austrian air is blowing. For it was through Karl Menger's work (professor at the University of Vienna since 1873, died in 1921, the same year as Dvořák) that the so-called Austrian School of Pure National Economy grew up, which undoubtedly ties in directly or indirectly with the theory of value judgement in Herbartianism; and its investigation has played a very important role in the Graz School of Meinong, in the von Ehrenfels School in Prague – where Zimmermann and Hanslick once came from – right up to the present day.[30]

What Schlosser notes here illustrates Landerer's point: the Herbartian theoretical moment in the Austro-Hungarian Empire is formalism, and here there is a singular conjunction in the formation of formalist theories in Vienna, a conjunction that takes hold from the 1850s onwards and lasts until the beginnings of the first republic.[31] One of the anchors is *Vom Musikalisch-Schönen* of 1854, a best-seller of the 19th century, whose links with Zimmermann's formalism are complex. Many years later, as Schlosser himself points out, Hans Kelsen's *Reine Rechtslehre* of 1934 is also understood as a formalistic program. What these programs have in common is a determined opposition to any kind of interpretation of content, the demand for an autonomous method, and the insistence on the objectivity of the results on which it is based; this formalism that seeks purity is not without political and social stakes, it is thought of as emancipation and control.

Above Schlosser mentions of course the school reform of the Ministry of Education of Count Leo von Thun, after 1848, a well-known episode which is at the heart of the spread of Herbartism.[32] This school reform will obviously be a key moment in the institutionalization of Herbartism, both at the university and at the high school, in Vienna, Prague and throughout the empire. From 1848 to 1853, this "Thunsche Reform"[33] was imposed in the Empire, led by an Herbartian team, originally from Prague and close to Bolzano: Count Leo von Thun-Hohenstein (1811–1888), an admirer and supporter of Bolzano even after his dismissal, Franz

---

29 Von Schlosser 1934, p. 191.

30 Von Schlosser 1934, p. 193. Research on Riegl often indirectly concerns the Herbartism: Vasold 2010.

31 Landerer 2009.

32 Stachel 1999, p. 135; Exner would have been the "architect", p. 141.

33 Stachel 1999, p. 135.

Exner (1802–1853),[34] Herbartian and a great personal friend of Bolzano, Robert Zimmermann, his intellectual son. This reform was inspired by Bolzano's Josephism.[35] It is necessary to insist on the strength of the personal ties that unite the protagonists of this Herbartian formalism, who develop an obvious tactic to occupy the philosophical field, a propensity that will come up against another tactic at the end of the century: the school of Brentano. The reform redefines the disciplines within the Vienna University, cutting out the field of university knowledge with striking effects,[36] it is a decisive breeding ground for the future Viennese school of art history.[37] It was under its aegis that the first chair of art history and archaeology was created in Vienna in 1852, to which Rudolf Eitelberger was appointed. It was also at this time that the first chair of musicology, entitled History and Aesthetics of Musical Art, was created, which Hanslick was awarded in 1861.[38] In the same year, Robert Zimmermann was elected to the chair of philosophy, one of the most powerful chairs in the empire because it was located in the heart of the empire. He held the chair for thirty-five years (1861–1896), and was practically the sole master on board for fifteen years after Brentano was deposed in 1880. The reform begins the publication of manuals that will be widely read and distributed throughout the empire. Zimmermann will participate strongly in the process: his *Propaedeutik* will be published and translated many times.[39] By founding the Austrian Museum of Art and Industry in 1864, Eitelberger gave a special dimension to the history of art in Austria-Hungary (where Riegl was curator of the textile section from 1887 to 1897).[40] A recent exhibition at the Belvedere in Vienna in March 2016, entitled "Kubismus, Konstruktivismus, Formkunst", emphasized

---

34 About Exner's dynasty: Coen 2007.

35 Johnston 1991, p. 328, considers this reform to be "the most enduring monument of Bohemian humanism". For Rampley, Thun's liberalism interprets Exner's initiatives conservatively (Rampley 2013: 16f.). It is worth noting Eitelberger's liberal commitment in 1848 and after, Rampley offering a very complete portrait of the latter which, however, says nothing about Zimmermann; see also Espagne 2011.

36 This ministry will also create a chair in the history of literature. Stachel insists on the need to specialize the disciplines, especially at the high-school level by abolishing what was customary, a single teacher for all subjects, and at the university level by creating a fully-fledged philosophy curriculum (Stachel 1999, p. 137f.).

37 Landerer 2009, p. 80: "l'école viennoise naît de l'esprit de l'herbartisme" and the Thun ministry baptized her...

38 Eitelberger's support of the Herbartians runs counter to Schlosser's image of him as a Hegelian, and is based on the primacy of the concrete object (von Schlosser 1934, p. 154–159).

39 Jäger 1982, p. 198, insists on dissemination of Herbartian aesthetic ideas through textbooks written by proponents of this school: Lindner, Drbal, and Zimmermann.

40 The Eitelberger case was recently analysed: Kernbauer et al. 2019.

the link between the strong intertwining of drawing education and the birth of abstraction in the pedagogical training of the empire.[41] Alexander Klee, rightly placing Robert Zimmermann at the forefront of his text, shows how the teaching of drawing in the monarchy was not only the transmission of a technique, but a critical relationship to the object, a constant reflection on the objectivity and objectuality of the perceived.[42] Illustrating his point with plates from textbooks of the time, Klee indicates the importance of the decomposition of mathematical forms that was inculcated, for example in Gustav Adolph Lindner's textbook, entitled in direct echo to Herbart, *Das ABC der Anschauung als Grundlage eines rationnellen Elementarunterrichtes im Zeichnen*:[43] intuition educates itself individually and collectively.

The reform of education, which took shape after the failure of the 1848 revolution, must guarantee the *status quo*, maintain national cohesion, and provide a "national philosophy" capable of cooling the revolutionary and separatist upsurge. Herbartism will play this role successfully. Against revolutionary aspirations, Herbartian political quietism reassures the authorities, its formalism devoid of any psychological anchorage promotes a supra-national identity beyond national particularities. Eitelberger was pleased that the Herbartian school never found itself in the midst of a confessional or political conflict.[44] This statement confirms the hypothesis of a prudent philosophy without theology, which each one appropriates as he wishes.[45] Vienna thus chooses a path of formal "purity" which must be understood from this political background.[46] One of the features of this reform, a feature that makes it possible to see it as a singular Austrian path, is its anti-Hegelianism. Franz Exner, in his *Psychologie der Hegelschen Schule* of 1842, displays "a hard fight" against the Hegelian school and thus against the

---

41 Maigné 2017, chapter 4.

42 Klee 2016, p. 17–33, here p. 18. On Gustav Adolf Lindner (1828–1887), see Zumr 2009, p. 141–151; for more details on other Herbartian aesthetics, see Schneider 2009, which deals with Friedrich Conrad Griepenkerl (1792–1849), Georg Eduard Bobrik (1802–1870), Josef Wilhelm Nahlowsky (1812–1885), Otto Flügel (1842–1914).

43 Lindner 1871. Lindner echoes Herbart, *Pestalozzi's Idee eines ABC der Anschauung untersucht und wissenschaftlich ausgeführt*, Göttingen 1802, a key text in our opinion regarding articulated decomposition and the need to exercise intuition (Maigné 2018), against Kantian transcendental aesthetics.

44 Landerer 2009, p. 77–79.

45 Jäger 1982, p. 199.

46 Landerer 2009, p. 74–76: this primacy of the "purity" of form, which is consistently found in various fields, in music (Hanslick but also Schönberg), in architecture (Loos), in law (Kelsen's Pure Theory of Law).

German school.[47] In their correspondence, Bolzano and Exner make fun of Hegel, are scandalized by his celebrity, and are pleased that they understand nothing of his writings: they see in them the sign of their own philosophical health, concerned with exactitude and clarity.[48] Hegelianism as a philosophy of history and thus emancipation in and through history is countered by a formal anti-psychologism (Herbart and Bolzano) and a scientific psychology (Herbart). In the course of the century, beyond the years 1850–1860, Herbartians will not escape the weight of nationalism: The controversy between Zimmermann and Hostinský is thus contemporary to the split in Prague of the Prague university into Karolinum (German) and Klementinum (Czech) in 1882, the opposition between Viennese "abstract formalism" and Prague "concrete formalism" having an obvious political resonance, the Viennese formalism being accused of rejecting a concreteness that calls into question supra-national abstraction.[49]

Austrian critical literature has often mentioned its double origin in Herbart and Bolzano: Winter already thought that the demand for accuracy in thought, against the idealistic *Schwärmerei*, explained the adequacy of Herbartian philosophy to the Austrian spirit.[50] Jan Sebestik defends the idea that the origin of the Austrian tradition lies *de facto* in Prague before being in Vienna.[51] In this context, Zimmermann's *Philosophische Propaedeutik*,[52] translated in particular into Hungarian and Polish,[53] which for several decades was to become a textbook of canonical philosophy throughout the monarchy, is also at the heart of the debates concerning the link between Herbartism and Bolzanism. The share of Bolzanian philoso-

47 Exner quoted by Clausberg 2011, p. 22. Clausberg develops the controversy that pitted Franz Exner against Karl Rosenkranz, p. 22–26.

48 The complicit alliance between Exner and Bolzano against idealism is, for example, evident in letters 5 (1833) and 6 (1834) of their correspondence. It is repeated again by Bolzano, who is delighted that Exner wrote his Über Nominalismus und Realismus against German idealism: "In his essay Über Nominalismus und Realismus, Professor Exner teaches us something that will certainly delight any friend of philosophy: he has long devoted his forces to defending solid boundaries and a secure position for formal logic, against the onslaughts of Hegel, Schleiermacher, and other not inconsiderable opponents." See Bolzano 2008 and 2017; Sebestik 1992, p. 121; Sebestik 1997, p. 38 and p. 57; Maigné 2008, p. 61.

49 Hostinský's texts frame this division: the one in 1877 (Das Musikalisch-Schöne und das Gesamtkunstwerk vom Standpunkt der formalen Aesthetik), the other one in 1891 (Herbarts Ästhetik).

50 Winter 1968, p. 167.

51 Sebestik 1992, p. 121; Sebestik 1997, p. 38 and p. 57.

52 Zimmermann 1853.

53 Barry Smith showed the importance of Zimmermann for the school in Lvov-Warsaw: Smith 1996, p. 155 ff.

phy within it was indeed open to discussion: the succession of editions indicates an increasingly strong Herbartian bias, gradually abandoning Bolzano;[54] if this is a reality, the Bolzanian filiation in aesthetics does not fade away with Zimmermann, on the contrary, it is fully integrated into it.[55] Watched over, isolated, Bolzano could not create a school, only a circle of intimates supported him, notably Franz Príhonský[56] and Michael Fesl; his texts were therefore poorly circulated and his theses were often put forward under the mask of Herbartism.[57] In 1883 Durdík wrote a significant article in the Herbartian *Zeitschrift für exakte Philosophie im Sinne des neueren philosophischen Realismus* on the spread of Herbartism in Bohemia, in which he stated that Bolzano had prepared the ground for Herbartian philosophy.[58] It is not surprising in this respect that the current Bolzanian studies constituted Zimmermann's first bibliography.[59] The question of psychology cannot be underestimated here: the quest for form and purity is accompanied by an innovative psychology in Herbart and his school, in particular by its mathematization and by the invention of the threshold of consciousness which dissociates representation and consciousness of representation: this psychology is very widely diffused in the empire thanks to Zimmermann's manual, all philosophers have read it.[60]

The importance of Prague in the history of Herbartism cannot be underestimated.[61] Franz Exner (1802–1853) influenced many scholars: Josef Dastich (1835–1870), Josef Nahlowsky (1812–1885), František Čupr (1822–1877), but also Josef Durdík's (1837–1902) and Otakar Hostinský (1847–1910). At the University of Prague, before, but even after, its division into Czech and German in 1882, the Durdík, Hostinský, Zich and Utitz chair of aesthetics followed one another. Durdík, professor of aesthetics in 1874,[62] wrote an important *Allgemeine Aesthetik* published in 1875, clearly from Herbartian perspective, but the homage to Bolzano is still alive[63]. In 1877 Hostinský's *Das Musikalisch-Schöne und das*

54 Morscher 1997 is debating with Winter 1975.

55 Maigné 2017.

56 F. Príhonský, one of the most talented disciples of Bolzano, wrote a famous text in 1850: *Neuer Anti-Kant*.

57 Johnston 1974; Blaukopf 1996.

58 Durdík 1883, p. 317–326.

59 Ganthaler/Neumaier 1997, p. 193–220.

60 Landerer/Huemer 2018. Die *Propaedeutik* is divided into two parts: *Empirische Psychologie* and *Formale Logik*.

61 Research of Tomáš Hlobil is important here, analyzing the period before 1848: cf. Hlobil 2012.

62 Neumann 1996, p. 114.

63 Durdík 1881.

Gesamtkunstwerk vom Standpunkt der formalen Ästhetik was written with Durdík's encouragement. Durdík and Hostinský thus have very close, personal ties, and are often regarded as the founders of musicology in Prague[64] (without forgetting Guido Adler, of course). They both established a filiation that would have repercussions as far back as the genesis of the Prague Circle[65] and into the work of Otakar Zich (1879–1934) and Emil Utitz (1883–1956).

Zimmermann's reign lasted thirty-five years (1861–1896) at the University of Vienna. Haller's thesis places this Viennese birth in 1874, the year in which Brentano's *Die Psychologie vom empirischen Standtpunkt* appeared.[66] Martin Seiler, continuing an analysis of Kurt Blaukopf, interprets the report of June 2, 1860, by Zimmermann's nomination commission for the chair of philosophy at the University of Vienna, as a "Manifesto of Austrian philosophy". It means a date of birth advancing its commonly accepted emergence with Brentano.[67] Zimmermann's actual appointment took place in January 1861. The argument of the three professors (H. Bonitz, F.K. Lott and F. Pfeiffer) is singularly Austrian, precisely according to the criteria put forward by Rudolf Haller, in that it engages a resolute criticism of the wanderings of idealism, especially Hegelian and Kantian, the concern of the authority of experience against the usurped power of the Idea, the importance of the method of natural sciences and mathematics, and in that it thus promotes a realism of Leibnizian inspiration. The commission wants to regenerate philosophy, in Vienna, against German philosophy and avoid the "Kantian interlude", according to the well-known expression of Otto Neurath, which however deserves to be nuanced, for we do not cease to discern the complex links with Kant within this tradition.[68] Against the catastrophe of German idealism, which dissolves all confidence in philosophy, the commission wants a philosopher professing a "realism", a "multiplied orientation in the fields of experience", supported by mathematics, against *a priori* speculation[69]. Zimmermann, in a way prolonging the reasons that got him elected, continuing the movement already

64 Neumann 1996, p. 120.

65 Stříteckÿ 1996.

66 Haller 1979. Haller identifies the following criteria: the rejection of Kant and Hegel, the promotion of Leibniz, gnoseological realism, the weight of the method of the natural sciences, the importance of the question of language; a philosophy that is both empirical and logical.

67 Seiler 2009.

68 Bonnet 2015.

69 Zimmermann is in concurrence with another Herbartian: Fridolin Wilhelm Volkmann (1822–1877), professor at Prague and author of a *Psychologie vom Standpunkte des philosophischen Realismus und nach genetischer Methode* (1856); several times reprinted and translated: Seiler 2009, p. 30.

mentioned in his history of aesthetics of 1858, gives a lecture *Über den Anteil Wiens an der deutschen Philosophie*, in 1886, where the specifically Austrian contribution is a definition of this philosophy: Zimmermann recalls the descendants of Leibniz in the empire, the reform of the teaching inspired by Josephism, anti-Kantianism and anti-Hegelianism, the importance of psychology.[70] The "Austrian" character of Austrian philosophy remains both astonishingly relevant and fragile, since each advanced trait (anti-psychologism, anti-idealism, realism, the weight of logic, anti-Kantianism, anti-Hegelianism, Leibnizianism) can find its opposite example, as the debates that immediately followed the use of this notion testify.[71] Otto Neurath had already made Bolzano the father of Austrian philosophy and estimated that it was thanks to him if the empire had avoided the Kantian interlude,[72] a statement which cannot mean that Kant's influence is absent in the empire. In spite of its difficulties of definition, there is indeed an "Austrian way towards modernity", and Herbartism is one key element of it, upstream of the Vienna Circle or "Vienna 1900".[73]

The issue of *Meinong Studies* that we present to the reader today would like to recall the historical and conceptual importance of Herbartism in the field of Austrian philosophy, by addressing several aspects, besides the field of aesthetics with which our introduction began: philosophical and theoretical, pedagogical, psychological. This volume hopes to be part of a currently active field, although too often fragmented between philosophical traditions that sometimes speak little to each other. It opens with Frederick Beiser's contribution on Herbart's realism, working on the very meaning of the term "realism" which was crucial for the Austrian tradition: neither naive nor indirect, Herbart's realism is a realism of the

---

70 Zimmermann 1886.

71 Haller 1979 already judged that it was impossible to circumscribe it definitively; Sauer 1982 thus insists on the equivocality of the criteria often used: if we consider the qualifiers "empiricist" and "realist", they are understood very differently by Brentano, Mach and the Vienna Circle, for example. Mulligan 2001 has shown how each criterion can be turned upside down to break down the barely constructed "Austrian" identity: if one considers gnoseological realism, Bolzano, Brentano and Meinong, for example, can certainly qualify, but an anti-realism marks the positions of Mach, Carnap and Boltzmann; if one takes Platonic realism, Bolzano and Meinong qualify, but not Brentano. The constants are in logic (theory of consequence, of probability), the taxonomy of acts and attitudes, the relation of representation to its object, and the question of intentionality, diversely understood, but which always poses the problem of the intuition. It is also necessary to underline the bearing of reflection on the emotions, in an approach that closely mixes logic, psychology, ethics and aesthetics, and from this point of view Herbart's formal aesthetics fully participates in this tradition.

72 Neurath 1981, p. 676. We also refer to the introduction of Bolzano 2010.

73 Herbartian tradition is already clearly identified in Jäger's seminal article 1982.

"giveness" where it is a question of studying the very appearance of things, a realism of relationships. Beiser shows the complex relationship of Herbart to Kant but also to Fichte. Exploring the precise links between Zimmermann and Brentano thus participates in an unprecedented historicization of the history of Austrian philosophy itself, as Denis Fisette's article on the Philosophical Society of the University of Vienna proves. The interest of working on this institution is to bring to light the strategies of the two camps face to face, Herbartians and Brentanians, at century's end where Herbartism is declining. Drawing on precise and abundant archives, Fisette shows what is at stake in the very term "science" to define philosophy, which is played out at several levels, the history of philosophy, psychology and the relationship to German philosophy. Riccardo Martinelli offers an in-depth analysis of the status of mental measurement, tracing a path of the relationship between psychology and psychophysics from Kant to Meinong, via Herbart, Hering and Brentano, a path that insists on finding Kant behind these Austrian debates, and where one can in a way "return to Kant" in an unexpected way. Katherine Arens offers a study of Herbart's pedagogy as epistemology, insisting on its communicational and embodied dimension, which closely links the state, the spirit of the individual and the naturalness of his body. She detects a future of this pedagogy between the Austro-Hungarian and American spaces, showing how the importance of forming the individual within a group can become a pedagogy of prescription and no longer of emancipation. Zumr proposes a Czech history of the history of Herbartism, in order to insist on an "Austrian" but not Viennese tradition, which the figure of Zimmermann could make us forget. He also insists on the fact that this tradition is rooted a long time before 1848, and explains the importance of Hegel's discussion within this tradition. Maigné insists on the cultural dimension of Zimmermann's *Formwissenschaft*, by studying the importance of the concept of *Phantasy* proposed in his mature texts. Each of the contributors nuances the outright judgments on realism, idealism, psychology, formalism, and thus discuss *de jure* and *de facto* the very concept of the Austrian tradition.

References

Beiser, Frederick (2014), *The Genesis of Neo-Kantianism 1796–1880*, Oxford: Oxford University Press.

Blaukopf, Kurt (1996), *Die Ästhetik B. Bolzanos*, Beiträge zur Bolzano-Forschung, Band 8, Sankt Augustin: Academia Verlag.

Bolzano, Bernard (2008), *De la méthode mathématique. Correspondance avec F. Exner*, C. Maigné/J. Sebestik (eds.), Paris: Vrin.

Bolzano, Bernard (2010), *Premiers écrits*, Paris: Vrin.

Bolzano, Bernard (2017), *Écrits esthétiques*, J. Sebestik/C. Maigné/N. Rialland (eds.), Paris: Vrin.

Bonnet, Christian (2015), "Kant en Autriche, rentre réception et rejet", in: *Philosophies autrichiennes, Austriaca* no. 78, Chr. Bonnet (ed.), Rouen: Presses universitaires de Rouen, p. 125–142.

Cassirer, Ernst (2000), *Das Erkenntnisproblem in der Philosophie und Wissenschaft der neueren Zeit*, Band III: *Die Nachkantischen Systeme*, Marcel Simon (ed.), in: *Ernst Cassirer Werke*, Band IV, Birgit Recki (ed.), Hamburg: Meiner.

Clausberg, Karl (2011), "'Wiener Schulen' im Rückblick. Eine kurze Bildergeschichte aus Kunst-, Natur- und Neurowissenschaften", in: E. Bisanz (ed.), *Das Bild zwischen Kognition und Kreativität, Interdisziplinäre Zugänge zum bildhaften Denken*, Bielefeld: Transcript, p. 21–73.

Coen, Deborah (2007), *Vienna in the Age of Uncertainty. Science, Liberalism and private Life*, Chicago: University of Chicago Press.

Durdík, Josef (1881), *O filosofii Bernarda Bolzana*, Akademický čtenářský spolek, Prague.

Durdík, Josef (1883), "Über die Verbreitung der Herbart'schen Philosophie im Böhmen", in: *Zeitschrift für exakte Philosophie im Sinne des neueren philosophischen Realismus* XII, p. 317–326.

Espagne, Michel (2011), "Rudolf Eitelberger von Edelberg 1817–1885 et les débuts de l'école viennoise", in: *L'école viennoise d'histoire de l'art, Austriaca* no. 72, C. Trautmann-Waller (ed.), Rouen: Presses universitaires de Rouen, p. 17–32.

Ganthaler, Heinrich/Neumaier, Otto (eds.) (1997), *Bolzano und die österreichische Geistesgeschichte, Beiträge zur Bolzano-Forschung* vol. 6, Sankt Augustin: Academia.

Haller, Rudolf (1979), *Studien zur Österreichischen Philosophie. Variationen über ein Thema*, Amsterdam: Rodopi.

Hegel, Georg Wilhelm Friedrich (1807), *Phänomenologie des Geistes*, Vorrede, reprint: *Werke in zwanzig Bänden*, E. Moldenhauer und K. M. Michel (eds.), Band 3, Frankfurt am Main: Suhrkamp 1971.

Henckmann, Wolfhardt (1985), "Probleme der allgemeinen Kunstwissenschaft", in: Lorenz Dittmann (ed.), *Kategorien und Methoden in der deutschen Kunstgeschichte, 1900–1930*, Stuttgart: Steiner, p. 273–334.

Herbart, Johann Friedrich, *Johann Friedrich Herbart's Sämtliche Werke in chronologischer Reihenfolge*, K. Kehrbach/O. Flügel (eds.), 19 vol., Langensalza 1887–1912. Reprint: Aalen: Sciencia Verlag (SW, following by volume).

Herbart, Johann Friedrich (1806, 1808^2), *Hauptpunkte der Metaphysik*, SW II.

Herbart, Johann Friedrich (1837), *Lehrbuch zur Einleitung in die Philosophie*, SW IV.

Herbart, Johann Friedrich (1828), *Allgemeine Metaphysik* I, SW VII.

Herbart, Johann Friedrich (1829), *Allgemeine Metaphysik* II, SW VIII.

Hlobil, Tomáš (2012), *Geschmacksbildung im Nationalinteresse. Die Anfänge der Prager Universitätsästhetik im mitteleuropäischen Kulturraum 1763–1805*, Hannover: Bochumer Quellen und Forschungen zum 18. Jahrhundert.

Hoeschen Andreas/Scheider Lothar (2001) (eds.), *Herbarts Kultursystem, Perspektiven der Transdiziplinarität*, Würzburg: Königshausen und Neumann.

Jäger, Georg (1982), "Die Herbartianische Ästhetik. Ein österreichischer Weg in die Moderne", in: H. Zeman (ed.), *Die österreichische Literatur, ihr Profil im 19. Jahrhundert (1830–1880)*, Graz: Akademische Druck- und Verlagsanstalt, p. 195–219.

Johnston, William M. (1991), *L'esprit viennois, une histoire intellectuelle et sociale 1848–1938*, tr. P-E. Dauzat, Paris: Presses universitaires de France.
Kernbauer, Eva et al. (eds.) (2019), *Rudolf Eitelberger von Edelberg. Netzwerk der Kunstwelt*, Vienna: Böhlau.
Klee, Alexander (2016), "Formkunst – Phänomen eines Kulturraums", in: *Kubismus, Konstruktivismus, Formkunst*, A. Husslein-Arco/A. Klee (eds.), Vienna: Belvedere Prestel, p. 17–33.
Koschnitzke, Rudolf (1988), *Herbart und Herbartschule*, Aalen: Scientia Verlag.
Landerer, Christoph (2005), "Die Geburt der Wiener Schule aus dem Geist des Herbartianismus", in: *Kunstgeschichte aktuell*, Jg. XXII, 2.
Landerer, Christoph (2004), *Eduard Hanslick und Bernard Bolzano. Ästhetisches Denken in Österreich in der Mitte des 19. Jahrhunderts*, St. Augustin: Academia, 2004.
Landerer, Christoph (1993), "Bernard Bolzano, Eduard Hanslick und die Geschichte des musikästhetischen Objektivismus", in: *Kriterion. Zeitschrift für Philosophie* V, p. 16–30.
Landerer, Christoph (2009), "Eduard Hanslick et le formalisme en Autriche: la voie apolitique de Vienne vers la modernité", in: Maigné/Trautmann-Waller (2009), p. 73–85.
Landerer, Christoph/Huemer, Wolfang (2018), "Johann Friedrich Herbart on Mind", in: S. Lapointe (ed.), *Philosophy of Mind in the Nineteenth Century* (Volume 5 of: *The History of the Philosophy of Mind*), London: Routledge, p. 60–76.
Lindner, Gustav Adolph (1871), "Das ABC der Anschauung als Grundlage eines rationellen Elementarunterrichtes im Zeichnen", in: *Programm des k.k. Gymnasiums zu Cili am Schluße des Schuljahres 1871*, Celje.
Maigné, Carole (2005), "Le réalisme rigoureux de J. F. Herbart" in: J. F. Herbart, *Points principaux de la Métaphysique*, traduction et édition française C. Maigné, Paris: Vrin, p. 7–162.
Maigné, Carole (2007), *Herbart*, Paris: Belin.
Maigné, Carole (2008), "Héritage bolzanien, héritage herbartien", in: B. Bolzano, *De la méthode mathématique. Correspondance avec F. Exner*, Paris: Vrin.
Maigné, Carole (ed.) (2012), *Formalisme esthétique. Prague et Vienne au XIXe siècle*, Paris: Vrin.
Maigné, Carole (2015) "L'esthétique autrichienne de l'école herbartienne", in: "Philosophies autrichiennes", *Austriaca* no. 78, Chr. Bonnet (ed.), Rouen: Presses universitaires de Rouen, p. 29–46.
Maigné, Carole (2017), *Une science autrichienne de la forme. Robert Zimmermann (1824–1898)*, Paris: Vrin.
Maigné, Carole (2018), "Herbarts ABC der Anschauung und der ästhetische Formalismus", in: *Herbart als Universitätslehrer*, J-F Goubet and R. Bolle (eds.), Jena: Paideia, p. 75-88.
Maigné, Carole/Trautmann-Waller, Céline (eds.) (2009), *Formalismes esthétiques et héritage herbartien. Vienne, Prague, Moscou*, Hildesheim/Zürich/New York: Olms.
Morscher Edgar (1997), "Robert Zimmermann – der Vermittler von Bolzanos Gedankengut? Zerstörung einer Legende", in: *Bolzano und die österreichische Geistesgeschichte, Beiträge zur Bolzano-Forschung* Bd. 6, Sankt Augustin: Academia Verlag.
Mulligan Kevin (2001) (ed.), *La philosophie autrichienne de Bolzano à Musil*, Paris: Vrin.
Neumann Kurt (1996), "Die *Allgemeine Ästhetik* von Josef Durdík. Ihr Einfluß auf die Genese einer positivistischen 'Schönheitswissenschaft' und auf die moderne Kunst", in: K. Blaukopf (ed.), *Philosophie, Literatur und Musik im Orchester der Wissenschaften*, Vienna: Hölder-Pichler-Tempsky, p. 106–130.

Neurath, Otto (1981), *Die Entwicklung des Wiener Kreises und die Zukunft des logischen Empirismus*, in: *Gesammelte philosophische und methodologische Schriften*, Bd. 2, R. Haller/H. Rutte (eds.), Vienna: Hölder-Pichler-Tempsky.

Orth, Ernst (2001), "Kultur und Vorstellungsmassen. Ansätze zur Entwicklung eines neuen Kulturbegriffs im 19. Jahrhundert bei Johann Friedrich Herbart", in: A. Hoeschen/L. Schneider (eds.), *Herbarts Kultursystem*, Würzburg: Königshausen und Neumann, p. 25–37.

Payzant, Geoffrey (2001), *Eduard Hanslick and Robert Zimmermann, a biographical sketch*, https://www.rodoni.ch/busoni/cronologia/Note/hanslick.pdf, visited on June 20, 2020.

Pettoello, Renato (2001), "Introduzione", J. F. Herbart, *Punti principali della metafisica*, tr. it. R. Pettoello, Turin: Thélème, p. VII–XX.

Pettoello, Renato (2002), "La realtà dell'apparenza ed i modi di dire l'essere. Il realismo senza oggetto di J. F. Herbart", in: *Le leggi del pensiero tra logica, ontologia e psicologia, il debattito austro-tedesco 1830–1930*, a cura di S. Poggi, Milan: Unicopli, p. 35–64.

Pettoello, Renato (2003), "Introduzione", in: J. F. Herbart, *Metafisica generale, con elementi di una teoria filosofia della nature, Parte systematica*, trad. it. R. Pettoello, Turin: Unione tipografico-editrice torinese, p. VII-XXXV.

Rampley, Matthew (2013), *The Vienna School of Art History. Empire and the Politics of Scholarship 1847–1918*, Pennsylvania States University Press.

Sauer, Werner (1982), *Österreichische Philosophie zwischen Aufklärung und Restauration. Beiträge zur Geschichte des Frühkantianismus in der Donaumonarchie*, Amsterdam: Rodopi.

Schlosser, Julius von (1934), "Die Wiener Schuler der Kunstgeschichte. Rückblick auf ein Säkulum deutscher Gelehrtenarbeit in Österreich", in: *Mitteilungen des Österreichischen Instituts für Geschichtsforschung*, Ergänzungsband XIII, H. 2, p. 141–228.

Schneider, Lothar (2009), "Quelques figures de l'esthétique herbartienne", in: Maigné/Trautmann-Waller (2009), p. 85–100.

Sebestik, Jan (1985), "Préhistoire du Cercle de Vienne", in: A. Soulez (ed.), *Manifeste du Cercle de Vienne et autres écrits*, Paris, Presses universitaires de France, p. 91–102.

Sebestik, Jan (1986), "Le Cercle de Vienne et ses sources autrichiennes", in: A. Soulez/J. Sebestik (eds.), *Le Cercle de Vienne. Doctrines et controverses*, Paris: Méridiens-Klincksieck.

Sebestik, Jan (1992), *Logique et mathématique chez Bernard Bolzano*, Paris: Vrin.

Sebestik, Jan (1994), "Prague Mosaic. Encounters with Prague Philosophers", in: *Axiomathes*, nuova serie, anno V, 2–3, p. 205–223.

Sebestik, Jan (1997), "Bolzano, Exner and the Origins of Analytical Philosophy", in: *Grazer Philosophische Studien* 53, p. 33–59.

Seiler, Martin (2009), "Un 'Manifeste de la philosophie autrichienne': la nomination du philosophe R. Zimmermann à l'université de Vienne (1860–1861)", in: Maigné/Trautmann-Waller (2009), p. 47–72.

Smith, Barry (1996), *Austrian Philosophy. The Legacy of Franz Brentano*, Chicago/La Salle, Ill.: Open Court.

Stachel, Peter (1999), "Das österreichische Bildungssystem zwischen 1749 und 1918", in: K. Acham (ed.), *Geschichte der österreichischen Humanwissenschaften*, vol. 1: *Historischer Kontext, wissenschaftssoziologische Befunde und methodologische Voraussetzungen*, Vienna: Passagen-Verlag, p. 115–146.

Stöckman, Ivo (ed.) (2019), *Texte der formalistischen Ästhetik, Eine Quellenedition zu Johann Friedrich Herbart und zur herbartianischen Theorietradition*, Berlin: De Gruyter.

Střítecký, Jaroslav (1996), "Vom Prager/Wiener Formalismus zum Prager Strukturalismus. Zu einer mitteleuropäischen Tradition", in: I. Bontinck (ed.), *Wege zu einer Wiener Schule der Musikssoziologie. Konvergenz der Disziplinen und empiristische Tradition*, Vienna: Guthmann & Peterson, p. 35–48.

Vasold, Georg (2010), "Alois Riegl und die Nationalökonomie", in: *Alois Riegl revisited. Beiträge zu Werk und Rezeption, Contributions to the Opus and its Reception*, P. Noever/A. Rosenauer/G. Vasold (ed.), Vienna: ÖAW, p. 29–36.

Venturi, Leo (1937), "Robert Zimmermann et les origines de la science de l'art", in: *Actes du deuxième congrès international d'esthétique et de science de l'art*, Paris: Alcan.

Winter, Eduard (1968), *Frühliberalismus in der Donaumonarchie. Religiöse, nationale und wissenschaftliche Strömungen von 1790–1868*, Berlin: Europa, p. 167.

Winter, Eduard (1975), "Robert Zimmermanns *Philosophische Propädeutik* und die Vorlagen aus der *Wissenschaftslehre* Bernard Bolzanos, eine Dokumentation zur Geschichte des Denkens und der Erziehung in der Donaumonarchie", in: *Sitzungsberichte der Österreichischen Akademie der Wissenschaften*, vol. 299, 5. Abhandlung.

Zimmermann, Robert (1853), *Philosophische Propaedeutik für Obergymnasien*, Vienna: Wilhelm Braumüller.

Zimmermann, Robert (1862), "Zur Reform der Aesthetik als exacter Wissenschaft", in: *Zeitschrift für exacte Philosophie* 2, H. 4, p. 309–358 (also in: Zimmermann, *Studien und Kritiken* I, Vienna: Braumüller 1870, p. 223–265).

Zimmermann, Robert (1886), *Über den Anteil Wiens an der deutschen Philosophie*, inaugural speech presented on October 14, 1886, Vienna: author's edition.

Zumr, Josef (2009), "Les écrits esthétiques de G. A. Lindner" in: Maigné/Trautmann-Waller (2009), p. 141–151.

Zumr, Josef (1998), *Máme-li kulturu, je naší vlast í Evropa. Herbartismus a česká filosofie*, Prague: Filosofia.

Frederick Beiser

Herbart's Realism

Abstract: This article is an examination of Herbart's alleged "realism". It argues that Herbart cannot be considered a realist in the usual modern senses of the term, because he denies that we have a direct or an inferential knowledge of reality itself. He insists that all that we directly know is our own representations, and he denies that we can have any knowledge of things-in-themselves. Herbart affirmed that he was a realist for the same reason that Kant denied that he was an idealist. Like Kant, Herbart affirmed the existence of things-in-themselves, and he disputed Berkeley's dictum *esse est percipi.* Nevertheless, Herbart's metaphysics makes two important departures from Kant's transcendental idealism: first, in allowing inferences beyond possible experience, and, second, in affirming that the relations between things are given in experience.

1 Really a Realist?

In his popular study of Herbart, Otto Flügel, a noted Herbart scholar, once wrote: "He who knows only a little about Herbart knows that he was a realist at a time when his age mostly thought idealistically."[1] Flügel treats Herbart's realism as if it were well-known what it meant, and as if there could not be any question that he held this doctrine. After all, he was only expressing a widely-held belief. In the early 20th century, when Flügel was writing, "realism" had become the catchword to describe Herbart's philosophy. Its characteristic doctrine was held to be its "realism", which supposedly distinguished it from all the idealist systems of his day.[2]

Whatever was known about Herbart's realism in the early 20th century, however, has not been passed down to us in the 21st. A survey of the literature of that time has revealed very little that explains, carefully and systematically, what

1 Flügel 1907, p. 1.
2 See Ueberweg 1902, p. 107–127.

Frederick Beiser, Syracuse University
fbeiser@syr.edu

https://doi.org/10.1515/9783110747324-002

Herbart means by realism.[3] But, even more problematically, there is nowhere in all his writings where Herbart himself defines, or explains in simple and straightforward terms, what he means by realism. We can reconstruct a tolerable sense of what Herbart means by "idealism";[4] but this still leaves indeterminate what he means by its polar opposite, realism. Hence the hapless historian has no choice but to reconstruct the meaning of the term from the various contexts in which Herbart uses it.

There can be no doubt *that* Herbart was a realist, at least in some sense of that very vague term. In a draft of a letter to C. A. Brandis, which was written in October 1831,[5] Herbart described his philosophy as "*mein Realismus*". What, exactly, he meant by his realism he did not explain. Still, if Herbart himself used the term, we have at least some reason to take it seriously as a description of his philosophy.

It is important to see, however, that Herbart did not use the term "realism" in any sense close to our contemporary senses of the word. Realism means, in most contemporary epistemology, the doctrine that our senses give us, in some form, knowledge of the world as it exists in itself, apart from our perception of it. Naïve or *direct* realism holds that the senses, *without the aid* of instruments or reasoning, give us such knowledge. Scientific or *indirect* realism maintains that the senses give us such knowledge only *with the aid* of instruments and reasoning. The senses serve only as the *material* of knowledge, the scientific realist says; they alone are not sufficient because we have to use instruments, measurement and reasoning to infer how reality exists independent of them.

If we use the word "realism" in either of these contemporary senses, then Herbart cannot be regarded as a realist. That he is not a naïve or direct realist is clear from the fact that he holds that the immediate objects of awareness are re-

---

3 There is no detailed discussion of Herbart in the various books about him. See the monographs by Flügel, Weiß, Franke, Fritzsch and Kinkel in the bibliography. The *Zeitschrift für exakte Philosophie*, the journal of the Herbartian school, which appeared in 20 volumes from 1861 to 1896 (suspended from 1876 to 1882), contained only one article treating Herbart's idealism or realism. This was M. W. Drobisch's article "Über die Wandlungen der Begriffe des Idealismus und Realismus und die idealistische Seite der Herbart'schen Metaphysik" (Drobisch 1865). Drobisch's article discusses more Herbart's idealism rather than realism, which he takes to be unproblematic.

4 See the fourth section of Herbart's *Allgemeine Metaphysik*, SW VIII, 198–244. All references to Herbart will be to *Sämtliche Werke*, eds. Karl Kehrbach and Otto Flügel, XIX vols., Langensalza: Hermann Beyer & Söhne 1887–1912. Reprinted Aalen: Scientia Verlag 1989. This edition will henceforth be abbreviated as SW. Roman numerals refer to volume numbers; Arabic numerals refer to page numbers. "§" designates a paragraph number.

5 Kehrbach attached the letter to his edition of the *Allgemeine Metaphysik*, Anhang III, SW VIII, 413.

presentations. We are not directly aware of objects themselves, Herbart maintains, but simply the representations or ideas that we have of them. In his *Einleitung in die Philosophie*, for example, Herbart teaches us that we are directly aware of only our own representations, and that for this reason we cannot have knowledge of objects outside us (§ 103, IV, 159f.). In a later polemical article, he writes that Schelling, in propounding a kind of realism, has forgotten one of the most basic lessons of Kant's critical philosophy:

> It is a well-known fact that, if there is to be any knowledge for us, we have it only *through our representations*; and that through all philosophizing we work immediately only upon our representations. Whoever forgets this and wants to jump into the real, falls into the old bog from which Kant tried to rescue his contemporaries...[6]

This principle – that we know directly only our own representations – runs so counter to realism that it was a mainstay of the idealist tradition. In his *Principles of Human Knowledge*, Berkeley had maintained that the immediate objects of perception are ideas, and on that basis he argued that we cannot have knowledge of any objects independent of them.[7] And in his *Kritik der reinen Vernunft*, Kant had stated in many passages that it is a fundamental principle of his transcendental idealism that appearances, i. e., what we perceive through the senses, are only representations.[8] Adopting this principle puts us more than half-way on the road to idealism. For the immediate implication of this principle is anti-realist: it implies that we cannot have any *direct* knowledge of reality itself, i. e., the world as it exists apart from and prior to our awareness of it. If we go on to claim that we cannot have any indirect or inferential knowledge of reality, then we have to abandon realism entirely.

Given Herbart's statements about our direct awareness of representations alone, it would seem that if he is a realist at all, he has to be an indirect or scientific realist. He must hold, in other words, that we know reality through inference or through measurement, instruments and reasoning. But it is very difficult to hold that Herbart is a realist in this sense too. He repudiated scientific realism no less than direct or naïve realism. It is a central thesis of scientific realism that we know the essence of things through their *mathematical* relations, that the objective properties of a thing are therefore its quantitative ones. But Herbart insisted, emphatically and expressly, that things in themselves are simple and indivisible, and that they are therefore unknowable by any quantitative or mathematical

6 "Über meinen Streit mit der Modephilosophie dieser Zeit", SW III, 325.
7 Berkeley 1949, §§ 4–6.
8 Kant, KrV, B 59, 66, 164, 235, 236, 518, 519, 527, 534, 535, 553, 554, 591.

method.[9] No measurements can describe their intrinsic natures, which are irreducibly qualitative.

That Herbart was far from realism in any contemporary sense of the word becomes plain when we recognize that, like Kant, he denies that human beings – whether through the senses or reason – have knowledge of things-in-themselves. It is a fundamental principle of his ontology that the ultimate units of reality are self-sufficient and independent; their intrinsic nature does not depend, therefore, on their relations to other things.[10] That intrinsic nature remains the same even when its relations with other things change. Since, however, we know things only through relations,[11] it follows that the ultimate units of reality are, in their intrinsic natures, unknowable. Herbart did not shirk, therefore, from explicitly embracing Kant's doctrine of the unknowability of things-in-themselves.[12]

At this point we are left wondering how Herbart can be a realist at all. He had denied naïve and scientific realism; and he affirmed the idealist principle that we know only our own representations. Why not, then, call it quits, dispute Herbart's self-ascription of realism and describe him as an idealist instead? Before we do that, though, we should go back in history and take account of Herbart's context.

2 Realism and Things-In-Themselves

In calling himself a realist, Herbart, a self-professed Kantian, was following a Kantian precedent. Foremost in his mind was Kant's famous dispute with Garve and Feder in the 1780's.[13] In 1784 Feder published in the *Zugaben zu den Göttingischen Anzeigen von gelehrten Sachen* a review by Christian Garve of Kant's *Kritik der reinen Vernunft*.[14] The review became notorious because in it Garve charged Kant with expounding a Berkeleyian idealism. Garve could see no difference between Kant and Berkeley: both held that the immediate objects of perception are ideas; and both denied that we could know the objective world through these ideas. In his *Prolegomena* Kant responded indignantly to Garve's charges.[15] He

---

9 *Allgemeine Metaphysik*, SW VIII, 66f., § 208.

10 I cannot embark on an account of Herbart's ontology here. I have done so in an earlier article, Beiser 2015, esp. p. 1064–68.

11 *Allgemeine Metaphysik*, SW VIII, 238, § 328.

12 *Hauptpuncte der Metaphysik*, SW II, 192, § 3.

13 On this dispute, see Beiser 1987, p. 172–177.

14 See *Zugaben zu den Göttingischen Anzeigen von gelehrten Sachen*, Januar 19, 1782, Stück 3, p. 40–48.

15 Kant 1903, p. 372–380.

repudiated any conflation of his transcendental idealism with Berkeley's idealism. Berkeley maintained that *esse est percipi,* that the very essence of things consists in their being perceived. But Kant put forward no such metaphysical principle; the aim of his philosophy was to limit metaphysical principles like Berkeley's, which claimed to know the essence of things or what reality is in itself. Kant protested that his philosophy cannot be Berkeleyian idealism because it taught that there is a realm of things-in-themselves, whose reality does not depend on our perception of them. Berkeley, Kant rightly emphasized, would never have permitted the existence of things-in-themselves. So, for Kant, the existence of things-in-themselves was one of the most important reasons his philosophy could not be described as idealism.

In describing his own philosophy as realism, Herbart was applying the same criterion that Kant once used in his dispute with Garve. His philosophy too was a species of realism, he maintained, because it affirmed the existence of things-in-themselves, a realm of things which exist apart from and prior to our consciousness of them. It was not a species of idealism, therefore, because it did not accept the principle *esse est percipi.* In a revealing passage from his *Hauptpuncte der Metaphysik* Herbart claims that Kant made a mistake in calling his philosophy a form of idealism.[16] Because his philosophy still presupposed the existence of things-in-themselves, Herbart implies, it would have been better described as a form of realism.

Here, then, lay one source of Herbart's realism: his commitment to the existence of things-in-themselves. This is certainly an unorthodox sense of the word "realism" by contemporary standards. But it is "realism" at least in the minimal sense that it contradicts one prevalent conception of idealism (namely, Berkeley's).

Herbart's commitment to the existence of things-in-themselves leaves us with a problem, however. Why did Herbart affirm this doctrine when it had proven such a source of difficulty for poor old Kant? Kant had stressed that all our knowledge is limited to experience; but he had also made clear that things-in-themselves do not lie within the boundaries of experience. How, then, do we know even of their existence? This difficulty was made notorious by Jacobi, who famously stated that without the thing-in-itself he could not enter Kant's philosophy, but that with it he could not remain inside it.[17] The same dilemma now seems to apply to Herbart, who, for the most part, affirmed Kant's limits on knowledge. Herbart regarded his own philosophy as a transcendental philosophy

---

16 SW II, § 10, 205.

17 See the "Beylage" to his *David Hume,* "Über den transcendentalen Idealismus" (Jacobi 1815), p. 291–310, esp. 304.

whose fundamental task is to know the limits of experience. He defined metaphysics as *"ars experientiam recti intelligendi"*[18]. Whence, then, the postulate of the thing-in-itself, which cannot be given in experience?

To answer this question, we have to take a close look at the sections on "Methodologie" and "Ontologie" in the *Allgemeine Metaphysik*. Following his definition of metaphysics, Herbart stressed that metaphysics has to begin with the given. It is only by taking account of, and beginning with the given, that metaphysics could avoid constructing castles in the air, postulating concepts that have no reference to anything that exists (§ 195, 49). The given is what separates reality from possibility, the real world from the world of fiction. But it was a special problem for him to determine *what* is in fact given. There is always the danger of reading or seeing too much in the given, so that one uses it to infer what one sets out to prove.

What is given to us, Herbart maintains, is the *matter* and *form* of sensation. The matter consists in the qualia of sensations; and the form consists in their relations to one another (§ 169, 19f.). These appear given to us because both the qualia and their relations appear to us independent of our will and imagination. We have no choice, and it is beyond our conscious control, that we perceive just these qualia and in just these relations to one another. Nevertheless, though the matter and form of sensation are given, Herbart cautions us from making any inferences from them. "Actual givenness, whether of the content or the constitution of a thing, cannot be considered a proof of existence." (§ 197, 51). This is because the skeptic will always dispute that there is anything corresponding in the real world to the content and form of sensation; he will maintain that they could well be nothing more than states of our consciousness. Of whatever is perceived in experience, the skeptic will be tempted to declare "You are nothing!" (§ 199, 53).

But Herbart thinks that there is a limit to such skepticism. Although we can doubt all the inferences made from sensations, we cannot doubt the existence of sensation itself. We still have to admit that there are appearances, and that these are at least something, that they are "a true proper not-nothing" (*ein recht eigentliches Nicht-Nichts*). If, however, we admit that appearances exist, then we also have to posit the existence of something that appears, something that is the source of these appearances (§ 199, 53). These appearances are indications of something that exists outside our consciousness, given that these appearances come and go independent of our will and imagination; we exert no conscious control over when and how they appear. Hence Herbart regards appearances as

18 Herbart, *Theoriae de attractione elementorum principia metaphysica*, § 1, in SW III, 160.

"indications" (*Hindeutungen*) of being. He coins the dictum: "So many appearances, so many indications of being" (§ 199, 53).

What is given in our experience, Herbart realizes, is more complicated than simple sensations. Sensations are not discretely separable from one another; they are interwoven and appear in groups (§ 201, 54f.). We seem to see *things* having many properties and not simply the properties floating on their own. The things are the *unity* of the properties, what makes them properties of one and the same thing. Herbart admits that it is difficult to conceive how one thing can have many properties; that is for him a central problem of metaphysics. But he still thinks that it is necessary to postulate such things. Only in that way can we make sense of our experience.

Hence Herbart postulates things-in-themselves. These things are substances, that which unifies a collection of properties, and of which they are predicated. He does not think that we can know these substances as they exist in themselves; we cannot explain how they unify their properties; and we cannot have knowledge of their intrinsic natures; but we are compelled to postulate their existence to explain the appearances of things, which consist in the fact that the matter and form of sensations are given, and that they are given as distinct collections of properties.

It is for all these reasons that Herbart introduces realism into his metaphysics. After these reflections, he declares in the "Methodologie" section of the *Allgemeine Metapysik:*

> The connections among our representations, so far as they are formed in experience, mirror certainly the connections between things among one another [...]; and this connection between what is in us and that which is outside us, is clear from psychology, so that from it there springs a not insignificant confirmation for the true realistic metaphysics.[19]

It is a mistake, Herbart warns us, to infer that the forms of experience derive from the original forms of the faculty of knowledge. That, he tells us, is "the false doctrine" of idealism.

A strict Kantian will question Herbart's argument on the grounds that we cannot use the principle of causality to make inferences beyond our experience. Herbart presupposes the principle of causality when he assumes that the substance unifies its properties, bringing them together into an intelligible whole, and when he assumes that the substance is the cause of sensations in the observer. In both cases he is taking the concept of cause beyond experience to infer something about the cause of experience, a transcendent application of the

19 SW II, § 170, 21.

principle of causality according to Kant. Herbart does not dispute that he is using the principle in a transcendent manner; but he does not see anything wrong with this in principle. He questions the whole Kantian analysis of the principle of causality, which makes it a rule for ordering the succession of appearances. This analysis of causality keeps it firmly within the realm of experience (i. e., the succession of perceptions); but this is for Herbart an unnecessary restriction of the principle. In his view, causality is a purely formal principle which connects the condition with the conditioned; it is completely irrelevant whether these terms appear in time. It is wrong to maintain with Hume and Kant, he argues, that the cause precedes and the effect follows because the condition-conditioned relationship need not be temporal at all. Indeed, cause and effect, strictly speaking, have to be simultaneous rather than successive, because "*A cause that does not yet work is not yet a cause!*"[20]

Herbart's critique of Hume's and Kant's analysis of causality is an interesting and important topic; but it will take us too far afield to examine it here. Suffice it to say that it is in virtue of Herbart's analysis of this concept that he believes that he is justified in applying it beyond experience to things-in-themselves.[21]

3 Realism and Relations

It would be a mistake to regard the existence of things-in-themselves as the sole basis for Herbart's realism. There was another important source of his realism, which we can appreciate as soon as we take account of Herbart's criticism of Kant's idealism.

It was a fundamental shortcoming of Kant's philosophy, Herbart charged, that he could not explain the determinate relations in which things appear to us in sense perception.[22] What particular relations appear, and when they appear, is independent of our will and imagination; we cannot do anything about them but must accept them as just given. That my book is on the desk, that Berlin is 387 miles from Munich, that Lincoln died before Grant —these are basic facts about our world which we cannot change. It is implausible to assume that they originate from the innate activity of the mind itself, because this would be to attribute more

---

20 *Psychologie als Wissenschaft*, § 142, Anmerkung; SW VI, 213. These are Herbart's italics.

21 The analysis of causality appears in *Psychologie als Wissenschaft*, § 142 and its long Anmerkung, SW VI, 202–222.

22 *Allgemeine Metaphysik*, § 320, SW VIII, 324.

than finite powers to it. After explaining Kant's views on sense perception, Herbart declared:

> The incorrectness [of his view] appears already with the question: Whence arise the determinate forms of determinate things? Whence the determinate temporal distances between particular perceptions? These questions are absolutely unanswerable according to the Kantian view.[23]

According to Herbart, Kant had no way to explain these facts on the principles of his idealism. This is because of Kant's fundamental distinction between the form and content of knowledge. While Kant assigned the matter of knowledge to sensibility, which receives given content, he allocated form to the understanding, which actively creates its representations.[24] It followed from this general classification that the forms, which consisted in relations, could not be given but had to be read into experience. Herbart thinks that this claim about relations is true on a very abstract level – for the categories of the understanding – but that it is false for particular things in our experience. Their *determinate* forms, he argues, are contingent for the understanding, i. e., it is possible for them to be replaced by other forms; any determinate form is still consistent with the general categories of the understanding. We cannot find any reason within the mind itself why it should create all the determinate forms of experience.

Herbart's differences with Kant become clearer by noting his critique of Kant's Transcendental Aesthetic. Kant maintains that all spatial and temporal relations are made possible by the a priori intuitions of space and time; particular relations are carved out of the general a priori intuitions, which precede and make possible the perceptions of particular spatial and temporal relations. Herbart maintains the contrary: that we first perceive determinate spatial and temporal relations and on that basis form the general abstraction of space and time; there are no a priori intuitions built into the mind, which is a *tabula rasa* at birth.[25] If we stick with the Kantian view that space and time arise solely from original intuitions in the mind, Herbart argues, then we cannot explain the determinate spatial and temporal relations which we find in experience.

---

23 *Psychologie als Wissenschaft*, § 144, SW VI, 226.

24 Kant, KrV, B 33f.

25 See *Allgemeine Metaphysik*, § 293, VIII, 187; *Psychologie als Wissenschaft*, §§ 109, 111, VI, 88, 91f.

Here, then, lies the other source of Herbart's realism. It consists in his contention that not all relations are created and the work of the mind; determinate relations are simply given, just as much as the content of sensations themselves.[26]

To understand this dimension of Herbart's realism, it is important to take note of his distinction between the matter and form of knowledge, which he makes in section IV of the *Allgemeine Metaphysik*. The matter of knowledge is sensation; and its form is its order, its organization into definite groups and series (§ 327, VIII, 237). Sensation is the beginning of all our knowledge, since it is through it that we learn of the *existence* of things; but we cannot learn from it the qualities of things themselves; the qualities of sensation depend entirely on our physiology (§ 327, 235). Although the matter of knowledge has no similarity with objects themselves, Herbart contends, we can still say this about its form (§ 327, VIII, 235). This form is no less given than the matter of sensation, because how sensations appear in relation to one another is independent of our conscious activity (§ 327, 237).

To say that we can know something about things through their forms is to make a fundamental claim about the cognitive worth of relations. Herbart does not hesitate to make just such a claim: "We live solely in relations, and need nothing further." (§ 328, 239). It is through relations that we know things, even if we cannot know from them anything about their relata (§ 328, 238).

> Hence our knowledge remains always formal; it forms relations, without knowing anything about the members of the relations; it proceeds from such a given that not the constitution of things but only their togetherness or non-togetherness is conceivable.[27]

Although Herbart insists that the relations between things are given, it is important to see that he does not think that these relations give us knowledge of things-in-themselves. The intrinsic nature of things-in-themselves, since it does not depend on relations, still remains unknowable. Nevertheless, things-in-themselves enter into relations with one another, and we can know them insofar as they do stand in such relations; these relations constitute the *appearances* of things-in-themselves. Hence what we know of things through their relations are only appearances of things-in-themselves. We have some knowledge of reality through appearances, Herbart insists, because these appearances are not only representations in us but they are appearances of objects themselves; they attach to the things of which they are appearances.

---

26 See *Allgemeine Metaphysik*, §§ 169–171, VIII, 19–22.

27 § 328, 238.

Herbart tried to give a more precise formulation of the knowledge we have of reality through relations with his concept of "objective appearance"[28]. The basic meaning of "objective appearance" is determined by its contrast with "subjective appearance". An objective appearance is ascribed to the object itself, whereas a subjective appearance arises from the subject alone (§ 292, 186). But how, exactly, or in what respects, does objective appearance arise from the object? An objective appearance arises from the object because it is the appearance of something given. What is given are sensations and the determinate relations between them. They are given because they do not depend on the conscious (and presumably subconscious) activity of the subject, its will and imagination. An objective appearance is how an object appears to a being with our physiology and senses; it depends as much on the object itself as on our sense organs and physiology. A subjective appearance, in contrast, is more like an illusion; it does not depend on the object at all but arises from the subject alone.

We need to note how Herbart distinguished his concept of objective appearance from Kant's own concept. Both Kant and Herbart claim that appearances are intersubjective, i. e., that they are universal and necessary because they must hold for all perceivers. But Herbart thinks that his concept of an appearance goes beyond Kant's in that it applies to something given, so that it is attached to something real. Herbart had two objections to Kant's concept of an appearance: 1) it is only a representation in consciousness, and therefore detached from something given or outside it; and 2) it holds only for all *human* perceivers but not for all subjective beings, human or non-human.[29]

4 Refutation of Idealism

Herbart developed his own realism in reaction against not only Kant's but also Fichte's idealism. When he attempted a refutation of idealism in his *Allgemeine Metaphysik* he had in mind specifically Fichte's idealism. "With Fichte, idealism stood at the summit; and from this it had to fall down and destroy itself."[30] From the failures of Fichte's idealism, Herbart believed, one could see the necessity of his own realism.

28 See chapter 4, "Vom objectiven Schein", in the third section of the *Allgemeine Metaphysik*, §§ 292–296, VIII, 186–191.

29 See Herbart's comments in *Allgemeine Metaphysik*, § 292, VIII, 186; and § 299, VIII, 195.

30 *Lehrbuch zur Einleitung in die Philosophie*, § 104, Anmerkung, IV, 162.

The refutation of idealism was a task of no small order, Herbart emphasized. Idealism was *prima facie* a very plausible position because it expressed the indubitable fact that my world is the world that I perceive.

> The world which appears to us is something perceived (*unser Wahrgenommenes*); therefore it is in us. The real world, from which we explain appearances, is also something thought; therefore it too is in us. According to this, we should make our own ego the basis of all that we perceive.[31]

Though the idealist began from a sound intuition, from a basic fact of our experience – that what we immediately know is only our representations – he drew false conclusions from it. It did not follow that the world is only what we perceive.

Herbart's critique of Fichte's idealism focuses mainly on Fichte's 1794 *Wissenschaftslehre,* which was his paradigm of idealism. Fichte's grand project in that work was to show how the basic features of reality – space, time, causality – derive from his first principle, which states that "the ego posits itself", i.e., the ego is that which makes itself. The basic structure of the world would then be shown to be a necessary condition for the self-production of the ego. Fichte admitted, however, that his deductive program ran into a basic problem: if the ego were to posit the world, then it had to posit something opposed to itself. The world as it appears to our senses is an obstacle to our activity, moral and mental, and so it should be characterized as a "non-ego" (*Nicht*-Ich). Throughout his first exposition of the *Wissenschaftslehre* Fichte tried to resolve this problem; he wanted to show that the ego, in positing the world, was still somehow involved in positing itself. But, for Herbart, all these efforts were in vain.[32] If the ego is that which is self-positing, it cannot, for just this reason, posit something opposed to itself, a non-ego. For how does that which is self-positing limit itself by positing something opposed to itself? How do we derive from a purely self-positing ego something self-oppositing? Fichte was demanding the impossible: that the self-positing ego destroy itself.

This was only the beginning of Fichte's problems, in Herbart's view. He argues that the self-positing ego not only cannot derive the non-ego, but that it also cannot be a sufficient basis to derive the variety of sense qualities of our experience.[33] What sense qualities appear to us, when they appear and how they appear, is strictly contingent for us and cannot be derived from his first principle of the self-positing ego. It is indifferent to the self-positing activity of the ego that it

31 *Lehrbuch zur Einleitung in die Philosophie*, § 104, Anmerkung, IV, 162.

32 See *Allgemeine Metaphysik*, §§ 323–325, VIII, 228–233.

33 See *Allgemeine Metaphysik*, § 169, VIII, 18–20; and § 327, VIII, 237.

confronts a world with one set of sense qualities rather than another; each is compatible with its self-positing activity. The specific properties that appear to our senses are simply given to us, and they occur in just this order rather than another completely independent of our will and imagination. This holds not only for the *matter* of experience, i. e., simple existence and quality of sensations, but also for its *form*, i. e., the order or connection between these sensations. When we see an apple, for example, as red, round and hard, we cannot substitute these properties with any others; what we see depends on factors beyond our conscious control. What we see is partly the result of things-in-themselves acting on our sensibility, and partly the result of our perceptive organs and activities, so that what we see are appearances of things-in-themselves.

Fichte himself was aware of the challenge of explaining the given sense qualities of experience from his idealist principles. Hence he made the desideratum of his idealism the explanation of "the feeling of necessity" behind our representations, i. e., the fact that they appear independent of our will and imagination.[34] Herbart's critique of Fichte's idealism is that he never really fulfilled this desideratum. Fichte saw the problem; but he refused to give up his idealism in the face of it. For Herbart, this was Fichte's fatal flaw. In the face of this problem he should have given up the principles of his idealism. This was indeed the motive for Herbart's own realism: this given manifold of sense qualities remained the chief stumbling block of idealism; and its existence was reason for pushing philosophy in the direction of realism.

Herbart's objections to idealism were not confined to the failures of its deduction program. He also found fatal problems with Fichte's fundamental concept, the concept of the ego. His first principle was supposed to express a completely transparent act of self-consciousness; to say that the ego is only what it posits itself to be is another way of saying that the ego is only what it knows itself to be; it is therefore pure self-consciousness or what Fichte called "subject-object identity". In his *Psychologie als Wissenschaft* Herbart subjected this principle to severe scrutiny.[35] Fichte's principle should state a completely self-evident and transparent act of self-awareness; the subject should know itself, and it should know that it knows itself. But, how, Herbart asks, does the ego know itself? It must reflect on itself to know itself; but this act of self-reflection presupposes a higher subject who is reflecting on itself; and how do we know that higher subject? This results in an infinite series of acts of self-reflection, so that we can never know the ego that reflects on itself. With great subtlety and sophistication,

34 Fichte 1845/46, p. 423f.

35 *Psychologie als Wissenschaft*, "Erster Abschnitt: Untersuchung über das Ich", SW V, 237–280.

Herbart develops several formulations of this kind of argument, though all of them come to the same result: that the ego forever eludes itself and remains opaque to self-reflection.

Now that we have seen what Herbart regarded as the chief difficulties of Kant's and Fichte's idealism it is easier to understand his own realism. Both Kant and Fichte had failed to explain the givenness and particularity of sense experience from their idealist principles. Realism responded to their difficulty by making a principle out of it; realism was for Herbart the affirmation of the givenness and particularity of experience in the face of its irreducibility to the activities of the ego. Ultimately, that is the proper sense which we should give to Herbart's "realism".

References

Beiser, Frederick (1987), *The Fate of Reason: German Philosophy from Kant to Fichte*. Cambridge, MA: Harvard University Press.

Beiser, Frederick (2015), "Herbart's Monadology", in: *British Journal for the History of Philosophy* 23 (6), p. 1056–1073.

Berkeley, George (1949), *Principles of Human Knowledge*, in: *The Works of George Berkeley*, vol. II, ed. T. E. Jessop. London: Thomas Nelson & Sons.

Drobisch, Moritz Wilhelm (1865), "Über die Wandlungen der Begriffe des Idealismus und Realismus und die idealistische Seite der Herbart'sche Metaphysik", in: *Zeitschrift für exakte Philosophie* V, p. 121–166.

Feder, Johann Friedrich (1782), „Kritik der reinen Vernunft", in: *Zugaben zu den Göttingischen Anzeigen gelehrter Sachen*, Januar 19, Stück 3, p. 40–48.

Fichte, Johann Gottlieb (1845/46), *Erste Einleitung in die Wissenschaftslehre*, in: *Werke*, ed. I. H. Fichte, vol. I, Berlin: Veit & Comp., p. 417–450.

Flügel, Otto (1907), *Herbarts Lehren und Leben*. Leipzig: Teubner.

Franke, Friedrich (1909), *J. F. Herbart. Grundzüge seiner Lehre*. Leipzig: G. J. Göschen.

Fritzsch, Theodor (1921), *Johann Friedrich Herbarts Leben und Lehre*. Leipzig: B. G. Teubner.

Herbart, Johann Friedrich (1887–1912), SW = *Sämtliche Werke*, eds. Karl Kehrbach and Otto Flügel, XIX vols., Langensalza: Hermann Beyer & Söhne. Reprinted Aalen: Scientia Verlag 1989.

Jacobi, Friedrich Heinrich (1815), "Über den transcendentalen Idealismus", in: *Werke*, vol. II, Leipzig: Fleischer, p. 291–310.

Kant, Immanuel (1903), *Prolegomena zu einer jeden zukünftigen Metaphysik*, in: *Kant's Gesammelte Schriften*, ed. Preußische Akademie der Wissenschaften, vol. IV, Berlin: Reimer.

Kant, Immanuel (1911), *Kritik der reinen Vernunft, 2nd ed. 1787*, in: *Kant's Gesammelte Schriften*, ed. Preußische Akademie der Wissenschaften, vol. III, Berlin: Reimer.

Kinkel, Walter (1903), *Johann Friedrich Herbart, sein Leben und seine Philosophie*, Gießen: J. Ricker.

Ueberweg, Friedrich (1902), *Grundriss der Geschichte der Philosophie*. Vierter Theil. *Das Neunzehnte Jahrhundert*, neunte Auflage, ed. Max Heinze. Berlin: Mittler und Sohn.

Weiß, Georg (1928), *Herbart und seine Schule*. Munich: Ernst Reinhardt.

Denis Fisette

Robert Zimmermann and Herbartianism in Vienna: The Critical Reception of Brentano and his Followers

Abstract: This study concerns an aspect of the reception of Herbartianism in Austria that has not been thoroughly investigated so far. It pertains to a controversy opposing Robert Zimmermann and Franz Brentano in the context of discussions that took place in the Philosophical Society of the University of Vienna. This study looks more specifically at three important episodes involving the Philosophical Society: first, the controversy over Herbartianism, second, that over the evaluation of Schelling's philosophy, and finally, the reception of Bolzano in Austria. I will first describe the circumstances that led Zimmermann to get involved in the Philosophical Society and the source of his controversy with Brentano and his followers. I will then comment on Zimmermann's address as chairman of the Philosophical Society and Brentano's reaction to Zimmermann's remarks on Schelling and the historical period to which he belongs. I will complete my analysis of Brentano's reaction with a summary of his evaluation of Herbart's philosophical program to which Zimmermann adhered. The last part focuses on Zimmermann's decisive role in the reception of Bolzano in Vienna in connection with the Bolzano Commission established by the Philosophical Society. I will conclude with brief remarks on Zimmermann's legacy in Vienna.

Introduction

In "My Last Wishes to Austria", written just before he left Vienna in 1895, Franz Brentano describes the state of philosophy in Austria when he arrived in Vienna in 1874:

> I came in a time when it had become completely clear about the emptiness of pompously inflated doctrinal systems, but where the seeds of true philosophy were still almost entirely

Note: Thanks to Carole Maigné for her helpful comments on an earlier draft of this paper.

Denis Fisette, University of Quebec at Montreal
fisette.denis@uqam.ca

https://doi.org/10.1515/9783110747324-003

> lacking. The minister Ausperg (Stremayr) believed that he found in me the man who was most suitable for bringing such a germ to Austria. I was called and I followed the call. I found the situation extremely sad: a Herbartian doctrine, but no Herbartian school (the hour had already passed for them); and that nothing was everything.[1]

Brentano undertook to implant the seed of an authentic philosophy through his program of philosophy as science, which he first exposed publicly in his inaugural address at the University of Vienna in 1874 (Brentano 1929a). Brentano's efforts were very successful considering that, after his departure from Vienna, most of the important chairs of philosophy in the Habsburg Empire were occupied by his students. But to achieve this goal, it was first necessary to dislodge Herbart's followers who, after the reform of 1849, held key positions in Austrian universities. Besides Brentano, who replaced Franz Karl Lott, an influential disciple of Herbart in Vienna (see Dahms, forthcoming; Dahms/Stadler 2015, p. 83–88), his student Carl Stumpf took over Wilhelm F. Volkmann's chair in 1879, another influential disciple of Herbart in Prague. The following year, Anton Marty obtained Johann Heinrich Loewe's chair, which marks the beginning of the school of Brentano in Prague that lasted until the late 1930s. Finally, Alois Höfler, a student of Brentano and Meinong, replaced two influential disciples of Herbart in Austria, namely Otto Willmann in Prague in 1903, and Theodor Vogt in Vienna in 1907.

One of the main proponents of Herbartianism in Austria, to whom Brentano refers in the excerpt above, is Robert von Zimmermann, who held a chair of philosophy at the University of Vienna from 1861 to 1896. Zimmermann began his studies in philosophy by attending Bolzano's lectures in Prague, and he then turned to Franz Serafin Exner, a student and supporter of Herbart, to supervise his dissertation in Prague in 1846. In 1849, Zimmermann was habilitated by Franz Karl Lott in Vienna, and in the same year, he inherited the extraordinary professor chair at Olmütz, a position he would keep until 1852, just before his appointment in Prague for a chair of ordinary professor. In 1861, he returned to Vienna where he was appointed ordinary professor. After Brentano's resignation from his chair in 1880, Zimmermann spent more than fifteen years in Vienna as the only full professor in the philosophy department and he therefore had to assume most of the administrative tasks. He also assumed the position of Rector of the University of Vienna during the 1866–1867 academic year, and he contributed to the foundation of two important societies in Vienna,

1 Brentano 1895a, p. 64. Brentano describes the situation in the same terms in a letter to Hugo Bergmann dated June 1, 1909 in which he adds the names of Franz K. Lott and Anton Günther (Brentano 1946, p. 125).

namely the Philosophical Society of the University of Vienna in 1888 and the Grillparzer Society the following year. After 35 years of loyal service in Vienna, he was appointed professor emeritus in 1896. He died on August 31, 1898 in his hometown of Prague.[2]

During the thirty years he spent in Vienna, Zimmermann taught philosophy to most students in the Faculty of Philosophy, and the influence he has had on some of them is due in part to the fact that he was, for a long period, the only examiner (*Prüfer*) in the philosophy department. He is known to have taught the composer Gustav Mahler and the physicist Ludwig Boltzmann, for example, and he had a great deal of influence on the art theorist and leader of the Vienna School of Art History, Alois Riegl (see Gubser 2006; Trautmann-Waller 2009; Wiesing 2016), as well as on the classical music theorist Eduard Hanslick (Zimmermann 1885; Blaukopf 1995, 2000; Payzant 2002). Most of Brentano's students, including Husserl[3] and K. Twardowski,[4] attended Zimmermann's lectures. Moreover, Zimmermann left us a rich and diversified contribution to several areas of philosophy, including the history of philosophy in Germany and Austria[5] and the field of aesthetics, in which he became known for his anti-idealist orientation (see Zimmermann, 1854) and his defence of aesthetic formalism inspired by Herbart.[6] Indeed, several aspects of Zimmermann's phil-

2 There are several biographies on Zimmermann, including these: *Jahresbericht* 11, 1897–1898; E. Reich 1899; H. Spitzer 1900.

3 In his mathematical curriculum from 1881 to 1883, the young Husserl studied philosophy in Vienna as a second discipline and it was then that he was examined in philosophy by Zimmermann and Vogt (see Rollinger 1999, p. 16f.). But as Brentano pointed out to Stumpf in a letter from October 18, 1886 on the occasion of his recommendation of Husserl to Stumpf in Halle, Husserl has in no way been influenced by Zimmermann (Brentano/Stumpf 2014, p. 260). However, P. Varga (2015, p. 101) claims that Zimmermann had for the young Husserl "die gleiche Bedeutung" as Brentano!

4 Twardowski (2017, p. 2) explicitly acknowledged Zimmermann's influence on his thought.

5 Let us note his marked interest, in his early publications and in three academic addresses in Prague, Olmütz, and Vienna, in the philosophy of Leibniz (Zimmermann 1847, 1849b, 1850, 1852b, 1861). Let us also mention his numerous studies published in the session reports of the Vienna Academy of Sciences, including those on Kant and Auguste Comte (Zimmermann 1886b, 1874) as well as his numerous reviews, from 1870 to 1898, of the German philosophical literature for the British journal *Athenaeum*.

6 See especially Zimmermann 1858, 1865. For a detailed exposition of Zimmermann's aesthetics and art history, and his discussions on the musical aesthetics of his time, see Blaukopf 1995, 1997, 2000; on Zimmermann's critique of the aesthetics of the Hegelians, see Zimmermann 1854; on his program of an aesthetic as science, see Zimmermann 1862.

osophy are known today thanks to the long-lasting influence of his work in the field of aesthetics and to recent studies on this important aspect of his work.[7]

This study concerns an aspect of the reception of Herbartianism in Austria that has not been thoroughly investigated so far. It pertains to a controversy opposing Robert Zimmermann and Franz Brentano in the context of discussions that took place in the Philosophical Society of the University of Vienna. I am mostly interested in three important episodes in the history of philosophy in Vienna that involve the Philosophical Society: first, the controversy over Herbartianism, second, that over the evaluation of Schelling's philosophy, and finally, the reception of Bolzano in Austria. I will first describe the circumstances that led Zimmermann to become involved in the Philosophical Society and the source of his controversy with Brentano and his followers. I will then comment on Zimmermann's address as chairman of the Philosophical Society and Brentano's reaction to Zimmermann's remarks on Schelling and the historical period to which he belongs. I will complete my analysis of Brentano's reaction with a summary of his evaluation of Herbart's philosophical program, to which Zimmermann adhered. The last part focuses on Zimmermann's decisive role in the reception of Bolzano in Vienna in connection with the Bolzano Commission created by the Philosophical Society. I will conclude with several brief remarks on Zimmermann's legacy in Vienna.

1 Zimmermann and the Philosophical Society of the University of Vienna

Let us first introduce this venerable institution which has been a privileged witness of the evolution of the history of philosophy in Austria and the theatre of many discussions, including that between Zimmermann and Brentano (Fisette 2014). The Philosophical Society of the University of Vienna owes its creation to Brentano's seminars taught to a large public composed of philosophers and non-philosophers of all horizons.[8] The circumstances surrounding this foundation are

7 There are many valuable works on Zimmermann's aesthetic, including the recent work of C. Maigné (2017) and some collective works on formalism in aesthetics: Maigné 2013; Maigné/Trautmann-Waller (eds.) 2009; see also Gubser 2006; Wiesing 2016; Moro 2009; Payzant 2002.

8 Alois Höfler, a philosophy student of Brentano and Meinong, provides further information about Brentano's lectures from which the Society originates (Höfler 1917). Another important testimony regarding the origins of the Philosophical Society is that of K. Twardowski in his autobiography, in which he mentions, in addition to Brentano's lectures, a reading group composed of Brentano's students, including Hans Schmidkunz, Alois Höfler, Christian von Ehrenfels, and

described in detail in several annual reports of the Society and by some of its members.[9] Although the names associated with most of its founding members are Brentano's students and, for the most part, sympathizers of Meinong's philosophy in Graz,[10] this organization would not have been recognized as a society of the University of Vienna without the support of several professors from the Faculty of Philosophy. This is because at that time Brentano's academic position as a lecturer in Vienna since 1880 and his tense relations with the ministry deprived him of all academic power and his support to this initiative was merely moral. This academic support came initially from two influential members of the Faculty of Philosophy, namely the psychiatrist Theodor Meynert (1833–1892), a close collaborator of Brentano and Freud, and one of the founding members of the Society,[11] and Zimmermann, who was in 1888 the only ordinary professor in the philosophy department, and who held the position of Rector of the University of Vienna during the academic year prior to the foundation of the Society. In several of the Society's annual reports, including the one marking its tenth anniversary, Zimmermann's contribution is highlighted:

> Since 1889, Zimmermann was chairman of the Philosophical Society, which owes him valuable advice and claims since its foundation. It was counsellor Zimmermann who, at the founding of the Society (in spring 1888), represented the interests and needs of the Philosophical Society in the high academic senate of the University of Vienna in the warmest and most convincing manner, and only thereby enabled to bear the title "Philosophical Society of the University of Vienna" more than just by name. [...] The connection between the Philosophical Society and the University, which was thus formally achieved, was a reference

Josef Kreibig who met regularly to discuss Aristotle's texts (Twardowski 2017, p. 5). These four philosophers were the most active members in the activities of the Society.

9 See in particular *Jahresbericht* 1912–1913, p. 3f.; 1888, p. 1).

10 But Meinong's name does not figure in these testimonies, and the day after the official foundation of the Society, Meinong wrote to Höfler: "I can hardly imagine that he could have silently invented a philosophical society without even letting me know a word about it." (Meinong an Höfler, 23.2.1888, Meinong-Nachlass, Karton LV, Nr. 4503). His friend Höfler quickly corrected this situation by adding Meinong's name to the list of the first members of the Society. Meinong maintained his membership in the Society until his death in 1918 (Dölling 1999, p. 74).

11 In a lecture delivered to the Philosophical Society on the occasion of Meynert's death, Höfler (1892, p. 6) emphasized the importance of his central contribution to this organization not only through his lectures and his active participation in the evening discussions, but also as a scientist interested in philosophical issues. Meynert's active participation in the Society had in fact a driving effect on several other scientists from the Faculty of Philosophy who decided to join the Society.

not to be underestimated for the organization of the membership and the scientific activities of the Society.[12]

During the Society's first year of existence, Alois Höfler presided mainly because of his competences in sciences and philosophy, but also because, like most of the other founding members of the Society, Höfler was closely related to Meinong. This student of Boltzmann and Stefan could thus serve as a mediator in the exchanges within the Philosophical Society between scientists and philosophers. But after only three semesters, Höfler resigned as chairman due to overwork (Höfler 1921, p. 10) and he was replaced by Zimmermann, who would chair the Philosophical Society from 1889 until his retirement in 1896.

In addition to presiding over the Society during this late period in his career, Zimmermann gave a few lectures, the most important of which, for our purposes, are that on the occasion of his appointment as chairman of the Society on November 16, 1889, and that on the occasion of his 70th birthday (Zimmermann 1893/1894, p. 5–8). He also delivered some lectures, the first of which, in 1889, on the beginnings of mathematical psychology in Vienna (Zimmermann 1889a), a second on the aesthetics of G. Semper (1893), and the third on Spinoza's politics (1895). His first lecture on the beginnings of mathematical psychology in Vienna is typical of his treatises on the history of philosophy. He claimed that Herbart was not the first, with his mathematical psychology, to have applied mathematics to psychology. Christian Wolff and the young Kant during his pre-critical period had already shown that one could quantify over the properties of mental phenomena. But Zimmermann was particularly interested in an Austrian precursor of mathematical psychology named Joseph Misley,[13] namely "because he belongs to Austria, and in the narrow sense, to Vienna itself, and yet or perhaps for that very reason has remained almost unknown by his real name" (Zimmermann 1889a, p. 3).[14] Herbart nevertheless remains for Zimmermann "the first and true instigator in the exact sense of mathematical psychology" (Zimmermann 1889a, p. 5).

---

12 *Jahresbericht* 1912–1913, p. 3. Compare with the first annual report in which Zimmermann is warmly thanked "for the effective representation of the Society's interests within the academic senate, and the high academic authorities for the trust which they granted to the newly founded Society by granting a space to the philosophical faculty" (*Jahresbericht* 1888, p. 6).

13 The Viennese Joseph Misley was the author of a book on the application of mathematics to purely mental objects (Misley 1818). This book was considerably expanded in the following years, the last edition dating from 1830.

14 By "his real name" because, as Zimmermann (1889a, p. 3) points out in this article, Ribot (1879, p. 35) misspelled his real name by calling him Riesley.

2 The Controversy over the Name of the Philosophical Society

Several indications show that there were overt tensions in Vienna between, on the one hand, Zimmermann as a supporter of Herbart's philosophy and, on the other hand, Brentano and his followers. The first clues can be found in two reports of the Philosophical Society which clearly indicate that the source of this conflict lay in Zimmermann's bias in favour of Herbart's philosophy and his endeavour to impose his views on the Philosophical Society.[15] In the first report, which highlights the 70th anniversary of Zimmermann's birth, the theologian L. Müllner, then Rector of the University of Vienna, discusses some rumours related to two major conflicts within the Philosophical Society, namely a bias towards Herbartianism and the status of philosophy as science:

> When the Philosophical Society of the University of Vienna was founded, many saw in it the creation of a party; and if that were the case, then the rumour would probably have been right that it was not the Herbart school that was supposed to rule it. They had in *their* hands the responsibility whether the young society would find a firm place at the first university of the empire: and you have carelessly considered sufficient to foster, without considering any other point of view and with the serious express intention, philosophy as science and only as such, in order to promote the granting of premises to the Society by the high academic senate.[16]

Müllner knew for sure that it was not merely a rumour, as confirmed by the Society's 25th-anniversary report, which emphasizes the monopoly of Herbart's philosophy since the reform of education. J. K. Kreibig, a student of Brentano and one of the authors of this report, welcomed the opportunity, twenty-five years after the foundation of the Society, to more freely discuss Zimmermann's peculiar attitude towards the Philosophical Society in general and philosophy in particular:

> Since the reform of high and middle schools around the year 1850, Herbartianism had become as much the official philosophy in Austria as Hegelianism had been in Prussia. However, in the decade when our Society was founded, such a monopoly had become more and more outdated. In the axis of this reversal were the person of Zimmermann, on the one hand, and Franz Brentano with his numerous pupils, on the other hand, as if they were two poles. Even though the independent thinking of those pupils had developed far beyond the doctrine of their teacher, so that there could no longer be any talk of a unified "Brentano School," they were nevertheless considered, from the outside – especially by Brentano's

15 On Zimmermann's program for a renewal of philosophy in Austria, see Seiler 2009.
16 *Jahresbericht* 1893/1894, p. 12.

> enlightened opponents – as forming a homogeneous group. Even if the first instigators and participants of the Society had never tried to give it a one-sided character – incidentally, adherents of completely different philosophies soon joined our circle from outside – Zimmermann had ample opportunity, simply through the composition of the membership, to use his full objectivity and also his good will. We may say today that this spirit of impartiality at once has remained the good scientific spirit of our Society.[17]

Kreibig claims that the motivation of Brentano's students in this conflict with Zimmermann was in no way to substitute one school for another, since, as this excerpt makes clear, there was, strictly speaking, no "Brentano school" in Vienna.[18] In addition, there were open conflicts between Brentano and Meinong, and between Meinong and Brentano's most orthodox students, namely Anton Marty in Prague, to name only a few.

One of the sources of this controversy lies in the name of the Philosophical Society, as Alfred Kastil, a student of Marty in Prague, later confirmed in one of the last lectures delivered before the Philosophical Society in 1936, under the title "Franz Brentanos Kritik der Antimetaphysiker", which is a reflection on the state of philosophy in Vienna nearly fifty years after the foundation of the Society (Kastil 2020). The first part of his talk focuses precisely on the Brentano-Zimmermann controversy and Brentano's lecture on Schelling (Brentano 1929c). Kastil confirms that what triggered this controversy lay in Zimmermann's reluctance to append the term "scientific" to the name of the Society[19] and the fact that he would have used all his authority, as Brentano confirms in his talk on Schelling, to make sure that the term "scientific" be banned forever from its program (Brentano 1929c, p. 131). Zimmermann's manoeuvre was indirectly intended against the philosophical program advocated by Brentano and his followers of philosophy as science

17 *Jahresbericht* 1912–1913, p. 6f.

18 Indeed, after his departure from Vienna in 1895, Brentano's name is mentioned nowhere in the Society's annual reports, and unlike most members of the Society, his death in 1917 was not even mentioned in the annual reports or in the meetings of the Society. Later, A. Kastil, a student of Marty in Prague, would deliver lectures on Brentano in the Philosophical Society. Nevertheless, Brentano's correspondence shows that he was aware of the activities of the Philosophical Society after 1895 and was aware of the controversy surrounding his succession in Vienna (Fisette 2014).

19 Kastil wrote: "Zimmermann was not as alone in that judgment as he seemed to think. And so he did not advertise in vain to recognize in these heroes a certain equality to the sober research of his time. It was suggested, in order not to give rise to the spirit of those a priori systematisers, to include the word 'scientific' in the title of the Society. It should be called the Society of Scientific Philosophy but his authority had enforced the rejection of the proposal, so that it remained simply a philosophical society." (Kastil 2020, p. 393).

(see Fisette 2021b), and we shall see that this controversy goes far beyond a semantic issue.

3 Zimmermann's Address on the Occasion of his 70th Birthday

The other source of this controversy is Zimmermann's inaugural address as chairman of the Society, delivered on November 16, 1889, (*Jahresbericht* 1889–1890, p. 2) to which Brentano reacted a month later in his talk on Schelling. Although Zimmermann's address was not published in the *Jahresberichte*, Brentano's critical remarks in his lecture on Schelling contain several references to Zimmermann's talk that allow for several overlaps with the other address delivered by Zimmermann some years later, again before the Philosophical Society, on the occasion of his 70th birthday (Zimmermann 1893/1894). Brentano's references also help establish several links with Zimmermann's historical studies published during the same period (Zimmermann 1888/1889, 1888, 1886a). This controversy relates more specifically to two divergent views on the history of philosophy in the nineteenth century, and it derives more specifically from Zimmermann's remarks in his address regarding the status and place of Schelling and Kantianism in the history of philosophy. This presumably constitutes Zimmermann's motivation in his opposition to the formulation of the title of the Society and that of Brentano in the choice of his conference's topic as he confirms in this quote: "I did not choose Schelling simply because our dear President [Zimmermann] has named him in particular but also because he represents most typically the philosophy of this bygone era" (Brentano 1929c, p. 105).

The subject of this dispute pertains to the evaluation of the period in the history of philosophy to which Schelling's philosophy belongs, and which Zimmermann describes as a golden age of philosophy, as shown by the following excerpt from Brentano's lecture:

> It is a fact that there cannot really be any stronger contrast than between great glory and deep contempt. And we understand the nostalgia well with which our esteemed President, at our last assembly, turned his eyes to the past time saying that the golden age of our German literature had also been that of the golden age of German philosophy, and he told us how much it affected him during his youth, and how he had still been able to look into Schelling's eyes, one of the epoch-making thinkers.[20]

20 Brentano 1929c, p. 104.

We shall see that, for Brentano at least, the stake of this controversy lies in his philosophy of history and his theory of the four phases in the history of philosophy.

That said, in this address, Zimmermann describes himself “as a living witness of half a century of the development of philosophy”:

> I participated in this change, I saw the last glow of the Hegelian school with my own eyes; I saw the break happen within the school. I saw with my own eyes the last hero of the heroic age of philosophy, Schelling. I saw him in his youth and his eyes still sparkling despite his old age.[21]

But Vienna and Austria are not appanages of Germany in terms of culture and philosophy when one considers

> that the city, which gave birth to a poet such as Grillparzer, a painter such as Schwind, composers like Schubert, gave a second home to Gluck, Haydn, Mozart, and Beethoven, and despite the circumstances, does not deserve any inglorious participation in the German science.[22]

However, Vienna’s contribution to German philosophy largely depends on the Viennese philosopher Karl Leonhard Reinhold,[23] to whom Zimmermann assigns a central place in the history of post-Kantian philosophy in Germany. In his inaugural address as Rector of the University of Vienna, which mainly stresses Reinhold’s contribution, Zimmermann (1886a, p. 40) even considers him “the second scholar of German philosophy”, the first being Kant. Kant’s philosophy obviously comes first because it represents the threshold of German philosophy:

---

21 *Jahresbericht* 1895–1896, p. 8. Compare with his inaugural address in Vienna (Zimmermann 1861, p. 5) in which Zimmermann says substantially the same thing.

22 Zimmermann 1886a, p. 28.

23 In his correspondence with Marty, Brentano wrote about Zimmermann’s inaugural address as Rector: “Zimmermann has recently held his inaugural address as Rector. But was it good? According to the reports I received – who was not present –, it seems doubtful. He talked about Vienna’s contributions to the history of philosophy and especially about a man who is not honored enough, a pure, sweet soul [Reinhold = rein hold]. And who was, he asked, this pure graceful (*Holde*)? It was Reinhold! At least Hartl, who made a lot of fun of this handy illustration of Zimmermann’s Aesthetic. Maybe it’s a fable, and we do well to wait for the print of the speech.” (Brentano, Brief an Marty, 28-10-86).

> Kant's philosophy is like the threshold on which you cannot stand still, but which one must have taken. Only the one who went through the gates of it is in the hallway of contemporary philosophy, which is still in the process of being built.[24]

Whatever Reinhold's merits and contribution to Kantianism and the history of German philosophy, his connection with Austrian philosophy is rather thin and it is questionable whether he is truly representative of Austria's contribution to the history of philosophy in the nineteenth century.

Nevertheless, Zimmermann's treatises on the history of philosophy in Austria contain a wealth of first-hand information on the main figures in the making of Austrian philosophy.[25] Among the important figures in Zimmermann's narrative, the most significant is undoubtedly Bolzano, who represented for the history of philosophy in Austria what Kant represented for German philosophy during many decades. Zimmermann acknowledges the major influence that Bolzano had on him and the history of philosophy in Austria (1893/1894, p. 7) and maintains that he is at the origin of a significant tradition in the history of philosophy in Austria. But in this work, at least, he merely evokes the memory of the author of *Wissenschaftslehre* who initiated him to philosophy:

> When we turn the focus towards the professorship here and in Prague in earlier times, then other pictures appear before our eyes, and Bolzano's venerable figure shines forth in the midst of his auditors full of admiration, dominating not only the minds, but also the souls; he had a major and lasting influence – the traces are still apparent today; but he was taken away from them because he was forced to resign from his professorship after a sixteen-year career, never to enter it again.[26]

That said, Zimmermann is best known as the most important advocate of Herbart's philosophy in nineteenth-century Austria.[27] As he repeatedly points out, one of the most significant moments in the development of philosophy in Austria was the reform of education in 1849 that the ministry entrusted to Count Leo Thun, who was also a close friend and the sponsor of Bolzano, and Franz

---

24 Zimmermann 1886a, p. 32. "Even though Kant's *Critique of Pure Reason* is perhaps the key to philosophy, it is not philosophy itself." (Ibid., p. 37). On the reception of Kant via Herbart, see Zimmermann 1882.

25 In his 1888 article "Wissenschaft und Literatur", which bears on the history of philosophy in Austria, Zimmermann briefly discusses most of the important figures in the history of philosophy in Austria, from Bolzano to his contemporaries.

26 Zimmermann 1893/1894, p. 7. See Zimmermann 1888/1889, p. 184–198, and 1849a for a detailed description of Bolzano's life and work.

27 On Herbart's influence on Zimmermann's philosophy, see Zimmermann (1871, 1872, 1873, 1876, 1877) and Bauer (1966).

Serafin Exner, a student and disciple of Herbart. The aim of this reform was to secularize philosophy in educational institutions and to implant Herbart's philosophy in Austria just as the Prussians had done a few years earlier with Hegel's philosophy. Zimmermann also saw in his master Exner the "founder of a Herbart school in Austria", and the

> counselor and friend of the leaders of the first Ministry of Education in Austria, Chr. v. Feuchtersleben and Count Leo Thun, together with likeminded men such as J. A. Zimmermann, Bonitz, Lott, who became the intellectual instigators of the radical reform of higher education.[28]

It is known that this reform also made it possible to disseminate Bolzano's ideas thanks, among other things, to Zimmermann's *Philosophical Propaedeutic* which, for several decades, became the canonical textbook of philosophy throughout the Habsburg Monarchy. In addition to Bolzano and Exner, Zimmermann also stresses the name of the theologian and philosopher Anton Günther, another student of Bolzano, who also exercised much influence in Austria (Zimmermann 1888, p. 146).[29]

Zimmermann did not pay much attention to contemporary philosophy in this address, but he claims elsewhere (Zimmermann, 1888/1889) that the main trend in philosophy at that time was decidedly empiricist, emphasizing in this regard the contribution of his colleagues from the Medical School

> where both the realistic school of Herbart and the Vienna School of Medicine, which was based on an empirical foundation, had paved the way, in a grandiose manner and more than anywhere else, to the same approach in the foundation and method of experience.[30]

On the other hand, Zimmermann also expresses his reservations regarding his colleagues' empiricist orientation:

> Philosophically trained Vienna naturalists such as Rokitansky, Stricker, Meynert among others, have set the tone: thinkers familiar and friendly with Comte's positive philosophy and the inductive method of the Englishman, such as the publisher and translator of John Stuart Mill, the learned interpreter of Herculean documents, Th. Gomperz, in logic, F. Brentano who

---

28 Zimmermann 1888, p. 146. However, in the fourth part of the study "Philosophie und Philosophen in Österreich" in which Zimmermann examines the outlines of the philosophy of his teacher Exner, he stresses again the importance of the reform of philosophy in Austria but denies this time that this reform was aimed at implanting the philosophy of Herbart in Austria (Zimmermann 1888/1889, p. 246). On Zimmermann and the reform of education, see Payzant 2002.

29 On A. Günther, see also Zimmermann (1888, p. 259–264) and Bauer (1966, p. 80–104).

30 Zimmermann 1888/1889, p. 268; Lesky 1976.

> has affinities of thought with A. Bain, in philosophy, the latter transplanted it in his pupils, the author of *Tonpsychologie*, C. Stumpf (now in Halle), A. Marty in Prague, A. Meinong in Graz, etc. A new, empirically-minded generation of young thinkers, perhaps more than wished for, seems to be developing, as their organ, and at the same time as a good sign of a free philosophical movement which can no longer be repressed in the future, and which most recently prevail at the University of Vienna on the model of the "Philosophical Society" in Berlin.[31]

This excerpt sheds new light on Zimmermann's reservations regarding the designation of the Philosophical Society as a "scientific" organization.

4 Brentano's Lecture on Schelling

Let us now turn to Brentano's reaction in his lecture on Schelling, which is actually composed of two different presentations dating from two different periods. The first is his probationary lecture for his habilitation at the University of Würzburg that he defended in July 1866; the second was delivered before the Philosophical Society on December 17, 1889 (Brentano 1929c). The circumstances surrounding the writing of the first version are well known: one of the jury members, Franz Hoffmann, a disciple of Schelling, had imposed that topic on Brentano's probationary conference entitled "On the main stages in the development of Schelling's philosophy and the scientific value of the last phase of his philosophy". The second lecture incorporates the text of the first version, which he uses this time in the context of his controversy with Zimmermann, to whom he refers in this talk without referencing him by name but by using his title of President of the Philosophical Society, and it pertains to the place of Schelling in the history of philosophy.[32]

As I pointed out above, the main stake of this controversy lies in Brentano's four-phase theory and his views on the past and future of philosophy. This theory is based on the idea that regularities that can be observed in the course of the history of philosophy since the Pre-Socratics obey a law according to which each of the three main periods in the history of philosophy evolves according to four distinct phases. The first phase is ascendant, and it is characterized by the philosophical orientation of philosophers such as Anaxagoras and the Ionian philosophy of nature, Aristotle and Plato in Antiquity, Alexander the Great and Thomas

31 Zimmermann 1888/1889, p. 269.

32 In their correspondence from 1889–1890, Brentano and Zimmermann also exchanged views on space, time, and causality.

Aquinas in the Middle Ages, or Descartes, Leibniz, and John Locke in modern philosophy. Brentano's theory is based on two criteria: the first is based on the method which, according to Brentano's fourth habilitation thesis, is the inductive method used in empirical sciences (Brentano 1895b). The second criterion is based on the primacy of theoretical over practical reason or, as Brentano puts it, a philosophy guided preferably by theoretical rather than practical interests. Philosophers who meet both of these criteria belong to an ascendant phase, like Aristotle, Aquinas, or Locke do, while the philosophers who depart from these criteria belong to one of the three declining phases in the history of philosophy. We are interested here in the phase of extreme decline which is called mysticism, and which is characterized by the invention of artificial means of knowledge acquisition and "a mystical elevation of intellectual life". It favours fantasies about facts. Its main advocates are Plotinus and Neoplatonism in Antiquity, Master Eckhart and Nicolas de Cues in the Middle Ages, and the partisans of German idealism in modern philosophy, including Schelling. Brentano's philosophy of history constitutes, somehow, a justification for his severe judgment on Schelling's philosophy as paradigmatic of philosophy's extreme decline.

In his lecture on Schelling, Brentano agrees with Zimmermann that Schelling's philosophy is constructed on fantasies and favours speculation over induction, thus moving philosophy away from science and towards the arts (Brentano 1929c, 125). He also agrees with Zimmermann that "the near future is just as certainly that of a philosophy of facts as Schelling's time was that of speculations and fantasy" (1929c, p. 123). However, Brentano's reservations concern Zimmermann's evaluation of this segment of the history of philosophy as the golden age of philosophy, particularly in the following quote from Zimmermann's address:

> When I last spoke at this place, gentlemen, the last veteran of that heroic philosophical time, the philosophical Proteus, was still alive, and, of fatality it seemed, was left out almost to the utmost limit of human existence, in order to show himself the changes in the most comprehensive train of thought of modern times. [...] I can still see him in front of me, the short man, with his eyes still ardent in old age and the powerful forehead, the sardonic smile on his lips over the changing time, which once idolized him as a youth and turned away from the man and almost mocked the old man as he moved from the faithful silent Munich inebriated with art to the doubting, noisy, sober, intellectual Berlin. It was the fate of philosophy itself that was depicted in Schelling: marvelled like a prophet, used and needed like obedient, persecuted and feared like a harmful instrument, finally laughed at and sidelined, like a brainless dreamer. That is then what happened to him [...]. It is worth investigating whether the philosophy itself

> is to blame for this aversion, or more likely it seems, a misguided orientation of philosophy.[33]

Note, however, that Zimmermann's evaluation of Schelling in this excerpt may surprise when one considers his sympathy for Herbart's philosophy and his critical positions towards Schelling in his writings on aesthetics, especially in his book *Schelling's Philosophie der Kunst* (Zimmermann 1875). The question indeed arises whether Zimmermann changed his positions in these late writings, a question which I cannot address in this study (see Maigné 2017).

In any case, Brentano clearly dissociates himself from Zimmermann's evaluation and he opposes his own program of philosophy as science, which is at the basis of his theory of the four phases (see Fisette 2021b). Brentano argues that the philosophical value of this program is immeasurably greater than that of speculative philosophy, which historically coincides with its decline (Brentano 1929c, p. 130). That is why Zimmermann is wrong in his characterization of Schelling's philosophy as the heroic time and the golden age of philosophy.

> It seems that I am here in strong opposition to our dear president [Zimmermann], whom I have been able to so often approve with all my heart. He called this time of philosophy the time of its glory, he named it the golden age of philosophy. But the opposition should be more apparent than real. Philosophy never shone as much as it did at the time, even if it was a superficial and ephemeral shine. And he had the right to use the expression "heroic time" if, as I would not doubt, it was for him to designate by that the immense personal gifts and the titanic power of the efforts of a man who once arose as well as the victories by which they subdued the world.[34]

However, after this modest concession, Brentano turns against Zimmermann the testimony of two of his heroes against the philosophy advocated by the idealists, starting with that of Herbart himself, who defended a form of realism in reaction to the kind of idealism advocated by Schelling, for example, and who once said about this kind of philosophy that it was totally lacking in the scientific sobriety required by a rigorous philosophy and that it had fallen "into the hands of an inebriated generation" (1929c, p. 130). The second testimony is that of the Austrian poet Grillparzer that Zimmermann evokes in his presidential address and in memory of whom he founded, with Emil Reich, the Grillparzer Society. Against Zimmermann, Brentano quoted Grillparzer's narrative of a meeting that he once had with Hegel: "I found Hegel as enjoyable, understandable, and conciliatory as his system then seemed *abstruse* and *intolerant*." And in one of his epigrams,

33 Zimmermann 1893/1894, p. 5f.

34 Brentano 1929c, p. 130.

he says, "I believe more readily in any miracle, than in such a system (the Hegelian system)" (1929c, p. 127).

Brentano concludes his lecture on Schelling by evoking the quarrel over the name of the Philosophical Society, and he strongly denounces the "authoritarian manoeuvres" by which the term "scientific" has been deleted from the Society's program; he warns against the direct or indirect influence that Schelling's and Hegel's offspring might exert on its orientation (1929c, p. 131). Brentano insists everywhere in his Vienna publications that the era of a priori construction of grand speculative systems is a bygone era and that the future of philosophy belongs only to philosophers engaged in a sublunary philosophy exercised in the spirit of empirical sciences.

5 Brentano and Herbart's Philosophical Program

Considering Brentano's criticism of Zimmermann's evaluation of Schelling's philosophy, the question arises as to Brentano's attitude towards Herbart's philosophical program, which represents one of the stakes of his controversy with Zimmermann. This is especially so since Brentano does not directly discuss Zimmermann's works, although he comments extensively on several aspects of Herbart's philosophy, especially in his lectures on practical philosophy, in which he discusses several aspects of Herbart's ethics (Brentano 1973).[35] Brentano also discusses several aspects of Herbart's psychology in *Psychology from an Empirical Standpoint*. But Brentano is much less critical of Herbart's ethics[36] than he is of his psychology. He claims that what is lacking above all in Herbart's psychology is a foundation in experience.[37] Brentano nonetheless welcomes Herbart's efforts

---

35 It is worth mentioning that the German version of Brentano's manuscript of these lectures has been much abridged by the German editor, and it contains more materials on Brentano's discussion with Herbart. See Ethikkolleg, MSS., Eth. 21 (p. 20563–20613), which is kept in the Houghton Library at Harvard University in Harvard.

36 In *The Origin of the Knowledge of Right and Wrong* (p. IX), Brentano says about Herbart's ethics: "Still his teaching remains in a certain aspect truly related with mine, while, on other sides, other celebrated attempts to discover a basis for ethics find in it points of contact" (see also p. 44).

37 Brentano ironically illustrates his criticism of Herbart through Goethe's poem entitled Cat-Pie: "A cook wants to fetch his own game in the forest, but knows little about wildlife; instead of bringing back a hare as booty, he goes home with a wild cat. He can apply all the refinements of the most elaborate cuisine: in vain!" (1895a, p. 36f.). He then quotes the last two stanzas of Goethe's poem:

The cat that's by the sportsman kill'd
No cook a hare can render.

to develop psychology as a science, although he denounces the arbitrary character of the principles of his mathematical psychology (Brentano 1995, p. 50). Brentano imputes some of Herbart's errors to the lack of experimentation in his research and the need for an institute of psychology (1895a, p. 36f.). Moreover, Brentano does not entirely reject the principles of Herbart's psychology, and in particular the equivalent of the fundamental principle of *Vorstellungsgrundlage*, which is clearly formulated in Herbart's *Psychologie als Wissenschaft* (§ 103). However, Brentano sharply criticizes Herbart's classification of mental phenomena based on the Kantian division between sensibility, intelligence, and will (Brentano 1995, p. 147), his conception of the categorical judgments conceived of, after Kant, as synthetic judgments, and he blames him for his substantialist conception of the mind (Brentano 1995, p. 127). Brentano (1995, p. 94) also discusses Herbart's treatment of consciousness and its unity in connection with Herbart's hypothesis of unconscious mental phenomena (Herbart 1824–1825, § 199). In this respect, Brentano opposes Herbart regarding one of the central theses of his *Psychology*, namely that all mental phenomena are conscious, and discusses, in this context, the infinite regress argument (1995, p. 78f.).

In his discussion of Herbart's ethics, Brentano is particularly concerned with two principles: on the one hand, Herbart's main thesis that ethics is a sub-discipline of aesthetics understood as a formal discipline (1995, p. 203), and on the other, the thesis that "the end is right if the efforts directed towards it are beautiful" (1995, p. 66). Brentano discusses this last thesis at length (1995, § 37, p. 75f.) and criticizes it for not satisfying criteria that Brentano imposes on the notion of "right end". Brentano claims that the issue is not addressed correctly in Herbart's theory: "Ought I necessarily to endeavor in a beautiful manner?" Since beauty is also, for Brentano, a question of feelings and appearance, beauty might constitute a motivation, but it cannot be decisive in the face of this question. True, Herbart distinguishes in his aesthetic the beautiful from the merely pleasant, and therefore between sensory and intellectual feeling or, in Herbart's own terms, between the content of a judgment of taste, which is purely theoretical, and the pleasure or displeasure provided by the contemplation of a work of art, for example. But why, then, asks Brentano, conceive in a purely theoretical and formal way the object that provides pleasure or displeasure to the judgment of taste? Herbart's answer is that the judgment of taste consists solely of relations constructed out of several elements, which taken individually are meaningless or indifferent in themselves. This is also the central thesis of Herbart's formalism advocated by Zimmermann, according to which relations alone determine beauty, whereas primary content and sense feelings are indifferent in this respect. Brentano sums up Herbart's theory as follows:

> Herbart thinks all beauty is based upon relations: each part of what, in combination, pleases or displeases, is indifferent, taken in itself. In music, for instance, no one of the separate tones, the relations between which form an interval – say a fifth or a third that is recognized in music –, has by itself anything of the character it takes on when they all sound together. Thus, he says, the matter is indifferent; only the form determines the judgment of taste.[38]

Brentano's main objection against Herbart's formalism is similar to the one on the basis of which he is opposed to British empiricists in his *Psychology from an Empirical Standpoint*. It consists in questioning the distinction between sense feelings, which have no object besides themselves, and intellectual emotions, which are intentional and have relations as their object. Brentano argues that it is not because mere feelings are not about relations that they have no object at all. That is why Brentano considers the Herbartian concept of judgment of taste to be contradictory insofar as it assimilates two very different things, namely moral and aesthetic taste, which, for Brentano, are matters of feelings, and feeling is not judging. It is in this sense that Herbart's judgment of taste transgresses Brentano's criteria (Brentano 1973, p. 75f.).

6 Zimmermann and the Reception of Bolzano in Austria

Another important aspect of Zimmermann's contribution to the history of philosophy in Austria lies in his role in the reception and transmission of Bolzano's ideas there. We saw that Zimmermann granted Bolzano a special status in the history of philosophy in nineteenth-century Austria, and we have emphasized that he himself contributed in several ways to the diffusion of Bolzano's ideas in the country. In this regard, several commentators of Zimmermann (Winter 1933, 1976; Morscher 1997; Künne 1999) have emphasized the importance of his *Philosophical Propaedeutic* as the textbook commissioned by the Ministry of Education after the education reform aimed at teaching the two main philosophical disciplines in most high schools in Austria, namely logic and psychology.[39] The first edition of the volume on logic was so influenced by Bolzano that some careful readers of Bolzano even accused him of plagiarism (Winter 1976; Morscher 1997). Be that as it may, it is through their philosophical training in Austrian high

38 Brentano 1973, p. 76.

39 Hence the division of Zimmermann's propaedeutic into two parts: the first deals with psychology and it was published in 1852, while the second is on logic and it was published the following year. It is the latter volume on logic whose content is largely inspired by Bolzano's *Wissenschaftslehre* (see Morscher, p. 161–165). However, the later editions of Zimmermann's *Propaedeutic* have been thoroughly reworked along the lines of Herbart's philosophy.

schools that several philosophers and scientists have been in touch with the basics of Bolzano's logic. For example, Twardowski (2017, p. 2), like most high school students in Austria, was introduced to philosophy via Zimmermann's *Propaedeutic.*[40] Most of Brentano's students[41] also significantly contributed to the dissemination of Bolzano's ideas: the most important are Benno Kerry (1885–1891), who influenced Twardowski (1894; 1999c, p. 24),[42] and Husserl, who, during the 1890s, in working manuscripts, lectures, and in *Logical Investigations*, specifically in relation to logical psychologism, paid much attention to Bolzano's philosophy.[43]

That being said, there is also a lesser-known aspect of Zimmermann's contribution to the reception of Bolzano in Austria, which is once again related to the Philosophical Society and which also involves Brentano's students. That episode begins with the rediscovery in 1903 of Bolzano's manuscripts, which Zimmermann inherited after Bolzano's death and which he had discreetly deposited at the library of the Vienna Academy of Sciences without mentioning their value,

40 Although the late version of Zimmermann's propaedeutic was severely criticized by Meinong and Höfler because of its Herbartian content (see Coen, 2007), it appears that Höfler's textbook, which replaced that of Zimmermann in the late 1880s, partially replicates the Bolzanian content of Zimmermann's textbook, which he integrates into Brentano's descriptive psychology framework (see Uebel 2000, p. 133). According to Uebel, Höfler's textbook would have exercised an influence on the Austrian members of the Vienna Circle comparable to that exercised by Zimmermann's *Propaedeutic* in circulating Bolzano's ideas (Uebel 2000, p. 109).

41 Brentano himself taught Bolzano's *Paradoxes of the Infinite* in his 1884–1885 lectures in Vienna, although, as his correspondence with H. Bergmann shows, he always deplored Bolzano's influence on his own students: "Daraufhin ist es nun geschehen, dass Meinong sowohl und Twardowski als Husserl und Kerry, der allerdings auch mehr von mir als von Zimmermann beeinflusst worden ist und sich, nachdem er schon lange Wien verlassen, noch in den letzten Jahren vor seinem fruehen Tod in brieflichen Verkehr mit mir setzte, in das Studium von Bolzano vertieften. [...] Aber die Verantwortung fuer so vieles Absonderliche und Absurde, wozu sowohl Meinong als Husserl unter Beruecksichtigung von Bolzano gelangt sind, darf ich doch vollstaendig ablehnen. Und wie gesagt, wie ich selbst von Bolzano nie auch nur einen einzigen Satz entnommen habe, so habe ich auch niemals meinen Schuelern glaubhaft gemacht, dass sie dort eine wahre Bereicherung ihrer philosophischen Erkenntnis gewinnen würden." (Brentano 1946, p. 125–126; Bergmann 1909, 1966).

42 At the very beginning of this book (1894, p. 15), Twardowski credits Bolzano and Zimmermann for the distinction between content and object of presentations.

43 Husserl 2001, 1994a, 1975. Husserl claims to have "rediscovered" Bolzano's philosophy (1975, p. 37).

especially of Bolzano's unpublished mathematical writings.[44] An excerpt from a report of the Philosophical Society dated from 1902–1903 summarizes the circumstances of this rediscovery by the members of the Society:[45]

> We still have to commemorate a bibliographical event, which could lead to a scientific achievement of outstanding importance for the Philosophical Society if everything goes as we wish thereafter. It was known to some members of the [...] Society that manuscripts from Bernhard Bolzano (1781 to 1848) were in the possession of the chairman and honorary president of the Philosophical Society, Dr. Robert von Zimmermann. Since this tradition almost threatened to break down, the board of the Society, at the beginning of this year, was keen to investigate the remaining of those precious manuscripts. After many unsuccessful attempts [...], a very comprehensive compilation of Bolzano's original manuscripts, whose content is partly philosophical but mainly mathematical, was in the Imperial Library. The Society's secretary, Robert von Sterneck, has subjected these unordered manuscripts to an examination, which already allows to reasonably estimate the size of the collection. For example, there was nothing less than a ready-to-print manuscript of a "theory of function" that testifies to a surprising degree of the actuality of his views. And these investigations, among others, are still unpublished fifty-five years after the author's death! Throughout this long period of time, however, until recently there were significant indications in Anton Marty's rectorate speech not to ignore the reminders to the effect that it would be worthwhile to give the outstanding thinker of Austria the only worthy monument by printing Bolzano's writings. Bolzano's relations with the longstanding chairman of the Society Zimmermann prompted the Philosophical Society to do everything in its power to contribute to the realization of that old wish cherished by so many.[46]

It was as a result of this discovery that Höfler and other members of the Society took steps to prepare the edition of Bolzano's manuscripts and to reissue his main works.[47] However, Höfler's project was delayed in part due to his moving to Prague in 1903.[48]

---

44 Morscher (1997, p. 180f.) calls into question Zimmermann's contribution to the dissemination of Bolzano's ideas, mainly because of his negligence in the management of this valuable heritage.

45 See also Kreibig 1914, p. 276; Höfler 1921, p. 10f.; Winter 1933, p. 218.

46 *Jahresbericht* 1902–1903, p. 6f.

47 It should be noted that well before the rediscovery of these manuscripts, Marty announced, in 1896, in his address as Rector of the German University in Prague, the establishment of a Bolzano Foundation in the German and the Czech University, whose primary mandate was to prepare a new edition of Bolzano's works (Marty 1916, p. 91; Künne 1997, p. 57). Hugo Bergmann, the author of an important book on Bolzano (Bergmann 1909) and a student of Brentano and Marty, also took steps to publish Bolzano's *Wissenschaftslehre* (see Künne 1999, p. 58).

48 Once in Prague, Höfler became a member of the *Deutsche Förderungsgesellschaft für Wissenschaft und Kunst*, which enabled him to obtain some of the financial support for his publication

The next important step in the reception of Bolzano in Vienna is the creation of a Bolzano Commission on March 9, 1914 by the members of the Philosophical Society. A few months earlier, Höfler presented to his colleagues and several members of the Society (*Jahresberichte* 1912–1913, p. 10) a draft of this commission project whose mandate was both to make known the works of the philosopher and to support the growing interest at that time for the father of Austrian philosophy.[49] Höfler highlighted the importance of the rediscovery of Bolzano's manuscripts in 1903 and the many initiatives undertaken by Brentano's students and some members of the Society to promote the innovative nature of Bolzano's thought and the value of his philosophical ideas:

> If I allowed myself to mention the establishment of a Bolzano Commission in the Philosophical Society, it is first and foremost to prevent that this interest in the great Austrian philosopher disappears and that it grows in a sustainable way [...] It is sufficient for the moment to emphasize that the Philosophical Society of the University of Vienna, by what it does for the work of Bolzano, now wishes to pay homage to him in the name of a philosophical society.[50]

In the year of the foundation of the Bolzano Commission, the Society published the first volume of *Wissenschaftslehre*[51] along with an article by Kreibig (1914) on the reception of Bolzano in Austria.[52] Kreibig argues that even though Bolzano's philosophy was ignored at the time by his contemporaries because they were "intoxicated by dialectics",[53] Bolzano remains the first Austrian thinker who deserves the name:

project (see Winter 1976, p. 29) despite Marty's opposition in the commission in charge of the evaluation of Höfler's application (see Marty's letter to Brentano from February 19, 1905; Gimpl 1999, p. 20f.).

49 According to Article 1 of the Statutes and Regulations of the Bolzano Commission, its main mandate was "Neudrücke der Werke Bernard Bolzanos zu veranstalten und die noch ungedruckten Schriften desselben herauszugeben".

50 *Jahresbericht* 1912–1913, p. 14.

51 Bolzano 1914. In 1920, the Society published Bolzano's *Paradoxien des Unendlichen* with H. Hahn's annotations (Bolzano 1920), and the four volumes of *Wissenschaftslehre* were reissued several years later by Höfler's student W. Schultz, who was also a member of the Bolzano Commission (Bolzano 1929–1931).

52 See also Kreibig (1905, p. 375f.), who insists on the importance of the Bolzanian distinction between content and object of presentation and its usage by Husserl, Höfler, Twardowski, Meinong, G. Uphues, the young Zimmermann, Kerry, and himself.

53 Kreibig offers an explanation of Bolzano's historical situation based on Brentano's four phases theory: "Bolzano war in Wahrheit um mehr als ein Menschenalter zu früh gekommen, um nach Verdienst gewürdigt zu werden. Seine Zeit war die des dialektischen Rausches, der

> So, it would seem that the history of philosophy in the last century on the soil of Austria was not a truly significant phenomenon unless it has a thinker in Bernard Bolzano, whose universality, depth, and acuteness deserve far more scientific admiration than so many glittering poetical thoughts of his deified contemporaries.[54]

To celebrate this occasion, one of the most loyal members of this organization since its founding, K. Twardowski, was invited to give a talk at the general meeting of the same year. Twardowski's two-part talk served as a basis for one of his most important philosophical treatises, "Actions and Products" (Twardowski 1999a), in which he conceives of intentional content on the model of Stumpf's *Gebilde* and Bolzano's *Sätze an sich*.[55] Twardowski's main concern in this talk is logical psychologism and Husserl's objections against Twardowski's psychologizing conception of meaning and intentional content in his 1894 book (Ingarden 1948, p. 28f. Twardowski's talk raised so much interest from the members of the Society that two further discussion sessions were added on this occasion.

7 Zimmermann's Legacy in Vienna

In 1895, Brentano resentfully left Austria, after having waited in vain for 15 years as a private lecturer, and despite the efforts of Zimmermann and his colleagues of the Faculty of philosophy for Brentano's reappointment as a full professor. Höfler considered the hiring of Mach to replace Brentano, and later, of Jodl, an affront to Brentano (Höfler 1917, p. 325). In fact, Zimmermann initially disagreed with the Minister's recommendation to grant Brentano's chair to Mach, but he later rallied by proposing a compromise which is summarized in this excerpt:

> In Ernst Mach one has an epistemologist based on inductive and experimental methods. With Jodl, there is now a representative of ethics and history of modern philosophy. In the classical philologist Theodor Gomperz [the father of the future professor of philosophy Heinrich Gomperz] one has a representative of ancient philosophy. It is now necessary to make room for Christian philosophy. In any case, Müllner, who was attended by many

blendenden Paradoxie, der mystischen Phrase. In solchen Epochen wird ein Denker von Bolzanos Art hochmütig ignoriert, wenn nicht verspottet." (Kreibig 1914, p. 287).

54 Kreibig 1914, p. 274.

55 Twardowski took over the main distinction between action (*Function*) and product (*Gebilde*) from another student of Brentano, Carl Stumpf (1906a, 1906b), who also significantly contributed to the reception of Bolzano in Germany. See Fisette 2021a.

> students of the faculty of philosophy at the theological faculty, should represent the philosophy of the Middle Ages.[56]

But this is not the only decision taken by the ministry after Brentano's departure that had direct consequences on the orientation of philosophy in Vienna. In 1896, Friedrich Jodl, known for his anticlerical positions, and his antidote, the theologian L. Müllner, were appointed in Vienna not only to fill Zimmermann's chair but also to counterbalance the hiring of Mach the previous year. Jodl was also known for his resolutely anti-Brentanian positions and his many manoeuvres aimed at breaking the monopoly of Brentano's students in Austrian universities.[57] As Höfler pointed out in his autobiography, Jodl's main mission, when he left Prague, was to eradicate from Austria this provincial clique of priests or former priests whom he called the Brentanoids:

> The philosophy at the German university is occupied and dominated in turn by the clericals, more precisely, "Brentanoide" or "Brentanote," as he calls them: the "clique" of a fashionable philosophy which retreats in its "provincial seclusion" – that is Jodl's view at any rate – with his rotten compromise between a "research without presupposition" and the deeply suspicious reactionary of a "liberal theology".[58]

Jodl's crusade against Brentano and his followers was carried out on several fronts, particularly in the Philosophical Society, which Jodl chaired after Höfler's departure for Prague in 1903 and until 1912 (Fisette 2014).

After Müllner's death in 1911 and Jodl's in 1913, the department appointed Robert Reininger as extraordinary professor in Vienna, and in 1922 as ordinary professor, the same year as K. Bühler and M. Schlick. Reininger was a student of Zimmermann who defended a dissertation on Schopenhauer in 1903 and was one of the few advocates of Kantianism in Vienna (see Nawratil 1998, 1969). Reininger became a member of the Philosophical Society at the turn of the twentieth century, under Jodl's presidency, with whom he was rather close, he became Vice President between 1906–1912, and after Höfler's death in 1922, he took over the chairmanship of the Philosophical Society until its dissolution in 1938.

Under Reininger's lead, the Philosophical Society underwent profound transformations which would later have as a consequence the distortion of the initial

56 Zimmermann, quoted in Wieser 1950, p. 39.

57 Jodl's career is closely related to Brentano's students. He moved to Prague in 1884 to replace C. Stumpf and became a colleague of Marty and Masaryk; in 1896, he was appointed in Vienna to fill Zimmermann's chair, and he was himself replaced in Prague by another student of Brentano, C. von Ehrenfels, who was also very active in the Philosophical Society.

58 Höfler 1892, p. 16; see Gimpl 1999.

vocation of the organization. Indeed, Reininger virtually abolished discussion time in the Society, which means, at the same time, its democratic character. He also reduced to a minimum the number of lectures delivered by scientists and non-philosophers, thus minimizing the interdisciplinary vocation of the Society, and abandoned the main projects dear to his predecessors, namely, that of the Bolzano Commission. In short, under his chairmanship, the society became just an organization among others, and it was quickly supplanted in Vienna by the Ernst Mach Verein, and later by the Vienna Circle. The final phase of the decline of the Philosophical Society was its annexation to the Kant-Gesellschaft, which Reininger celebrated on November 18, 1927.[59]

References

Jahresbericht der philosophischen Gesellschaft an der Universität zu Wien, Vienna: author's edition 1888–1918.

Wissenschaftliche Beilage zum Jahresbericht der philosophischen Gesellschaft an der Universität zu Wien, Vienna: author's edition 1902–1911.

Jahrbuch der philosophischen Gesellschaft an der Universität zu Wien, Vienna: author's edition 1912–1916.

Wissenschaftliche Jahresberichte, Vienna: author's edition 1926–1935.

Bericht, in: *Kant-Studien* 22, 1927, p. 556.

Bauer, Roger (1966), *Der Idealismus und seine Gegner in Österreich*, Heidelberg: Carl Winter.

Bergmann, Hugo (1966), "Brentano und Bolzano", in: *Archiv für Geschichte der Philosophie* 48, p. 306–311.

Bergmann, Hugo (1909), *Das philosophische Werk Bernard Bolzanos*, Halle: Niemeyer.

Blaukopf, Kurt (1995), *Pioniere empiristischer Musikforschung. Österreich und Böhmen als Wiege der modernen Kunstsoziologie*, Vienna: Hölder-Pichler-Tempsky.

Blaukopf, Kurt (1997), "Im Geiste Bolzanos und Herbarts. Ansätze empiristischer Musikforschung in Wien und Prag", in: H. Ganthaler/O. Neumaier (eds.), *Bolzano und die österreichische Geistesgeschichte*, Sankt Augustin: Academia Verlag, p. 237–278.

Blaukopf, Kurt (2000), "Von der Ästhetik zur 'Zweigwissenschaft'. Robert Zimmermann als Vorläufer des Wiener Kreises", in: M. Seiler/F. Stadler (eds.), *Kunst, Kunsttheorie und Kunstforschung im wissenschaftlichen Diskurs*, Vienna: ÖBV & HPT, p. 35–48.

Bolzano, Bernard (1914), *Wissenschaftslehre*, Bd. I, hrsg. von A. Höfler, Leipzig: F. Meiner.

59 "On November 18, 1927 was held, under the presidency of Professor Robert Reininger in Vienna, the general meeting of the 'Philosophical Society' during which the 'Philosophical Society', to the request of the commission, was acknowledged as a local group of the Kantian Society. From now on, it will bear the title of 'Philosophical Society of the University of Vienna', local group of the Kantian Society in Vienna." (Bericht 1927, p. 556).

Bolzano, Bernard (1920), *Paradoxien des Unendlichen*, hrsg. von A. Höfler, mit Anmerkungen von H. Hahn, Leipzig: F. Meiner.
Bolzano, Bernard (1929), *Wissenschaftslehre*, 4 Bde., hrsg. von W. Schultz, Leipzig: F. Meiner.
Brentano, Franz (1869), "Auguste Comte und die positive Philosophie", in: *Chilianeum. Blätter für katholische Wissenschaft, Kunst und Leben* 2, p. 15–37.
Brentano, Franz (1895a), *Meine letzten Wünsche für Österreich*, Stuttgart: Cotta.
Brentano, Franz (1895b), "Die vier Phasen der Philosophie und ihr augenblicklicher Stand", in: O. Kraus (ed.), *Die vier Phasen der Philosophie und ihr augenblicklicher Stand*, Stuttgart: Cotta, p. 1–31; trans. *The Four Phases of Philosophy*, M. Balsz and B. Smith (eds.), Amsterdam: Rodopi, 1998.
Brentano, Franz (1902), *The Origin of the Knowledge of Right and Wrong*, trans. C. Hague, Westminster: Archibald Constable & Co.
Brentano, Franz (1929a), "Über die Gründe der Entmutigung auf philosophischem Gebiet", in: O. Kraus (ed.), *Über die Zukunft der Philosophie*, Hamburg: Meiner, p. 82–100.
Brentano, Franz (1929b), "Über die Zukunft der Philosophie", in O. Kraus (ed.), *Über die Zukunft der Philosophie*, Hamburg: Meiner, p. 1–48.
Brentano, Franz (1929c), "Über Schellings Philosophie", in O. Kraus (ed.), *Über die Zukunft der Philosophie*, Hamburg: Meiner, p. 103–132.
Brentano, Franz (1946), "Briefe Franz Brentanos an Hugo Bergmann", in: *Philosophy and Phenomenological Research* 7, p. 83–158.
Brentano, Franz (1973), *The Foundation and Construction of Ethics*, trans. E. H. Schneewind, New York: Routledge.
Brentano, Franz (1995), *Psychology from an Empirical Standpoint*, trans. A. C. Rancurello, D. B. Terrell, L. McAlister, London: Routledge.
Brentano, Franz (2008), *Psychologie vom empirischen Standpunkte* (1874), M. Antonelli (ed.), *Schriften* I.1, Frankfurt: Ontos Verlag.
Brentano, Franz/Stumpf, Carl (2014), *Briefwechsel 1867–1917*, T. Binder et al. (eds.), Frankfurt: P. Lang.
Coen, Deborah R. (2007), *Vienna in the Age of Uncertainty. Science, Liberalism, and Private Life*, Chicago: University of Chicago Press.
Dahms, Hans-Joachim (2020), "Brentano's Appointment in Vienna", in: D. Fisette/G. Fréchette/F. Stadler (eds.), *Franz Brentano and Austrian Philosophy*, Vienna Circle Institute Yearbook, Berlin: Springer, p. 117–134.
Dahms, Hans-Joachim/Stadler, Friedrich (2015), "Die Philosophie an der Universität Wien von 1848 bis zur Gegenwart", in K. Kniefacz/E. Nemeth/H. Posch/F. Stadler (eds.), *650 Jahre Universität Wien – Aufbruch ins neue Jahrhundert*, Vienna: Vienna University Press, p. 77–131.
Dölling, Evelyn (1999), *"Wahrheit Suchen und Wahrheit Bekennen." Alexius Meinong: Skizze seines Lebens*, Amsterdam: Rodopi.
Eder, Gabriele J. (ed.) (1995), *Alexius Meinong und Guido Adler. Eine Freundschaft in Briefen*, Amsterdam: Rodopi.
Fisette, Denis (2014), "Austrian Philosophy and its Institutions: Remarks on the Philosophical Society of the University of Vienna (1888–1938)", in: A. Reboul (ed.), *Philosophical Papers Dedicated to Kevin Mulligan*, Berlin: Springer, p. 349–374.
Fisette, Denis (2021a), "Overcoming Psychologism. Twardowski on Actions and Products", in: A. Dewalque/C. Gauvry/S. Richard (eds.), *Philosophy of Language in the Brentanian Tradition*, Basingstoke: Palgrave, p. 189–205.

Fisette, Denis (2021b), "Remarks on the Architecture of Brentano's Philosophical Program", in: T. Binder/M. Antonelli (eds.), *The Philosophy of Franz Brentano*, Amsterdam: Brill, p. 28–49.

Gimpl, Georg (1999), "Philosophie und Interesse? Bernard Bolzano im Tauziehen nationaler Anspruchnahmen", in: *Nachrichten. Forschungsstelle und Dokumentationszentrum für österreichische Philosophie* 9, p. 19–46.

Gubser, Mike (2006), *Time's Visible Surface. Alois Riegl and the Discourse on History and Temporality in Fin-de-Siècle Vienna*, Detroit: Wayne State University Press.

Herbart, Johann Friedrich (1824–1825), *Psychologie als Wissenschaft: neu gegründet auf Erfahrung, Metaphysik und Mathematik*, vol. I u. II, Königsberg: A. W. Unzer.

Höfler, Alois (1892), *Worte der Erinnerung an Theodor Meynert und an sein Verhältnis zur philosophischen Gesellschaft an der Universität zu Wien*, Vienna: Braumüller.

Höfler, Alois (1917), "Franz Brentano in Wien", in: *Süddeutsche Monatshefte* 4, p. 319–325.

Höfler, Alois (1921), "Die Philosophie des Alois Höfler", in R. Schmidt (ed.), *Die deutsche Philosophie der Gegenwart in Selbstdarstellungen*, vol. II, Leipzig: Meiner, p. 117–160.

Husserl, Edmund (1975), *A Draft of a Preface to the Logical Investigations*, edited by E. Fink and P. J. Bossert, trans. P. J. Bossert and C. H. Peters, The Hague: Martinus Nijhoff.

Husserl, Edmund (1982), *Logical Investigations*, vol. 1, transl. J. N. Findlay, 2nd ed., London: Routledge.

Husserl, Edmund (1990–1991), "Ursprüngliche Druckfassung der Abhandlung 'Intentionale Gegenstände von Husserl'", in: Karl Schuhmann, "Husserls Abhandlung 'intentionale Gegenstände'. Edition der ursprünglichen Druckfassung", *Brentano Studien* III, p. 142–176.

Husserl, Edmund (1994a) "Critical Discussion of K. Twardowski, *Zur Lehre vom Inhalt und Gegenstand der Vorstellungen. Eine psychologische Untersuchung*", in: Husserl, *Early Writings in the Philosophy of Logic and Mathematics*, Berlin: Springer, p. 388–395.

Husserl, Edmund (1994b), *Briefwechsel*, 10 vols., vol. I: *Die Brentanoschule*, K. & E. Schuhmann (eds.), Dordrecht: Kluwer Academic Publishers.

Husserl, Edmund (2001a), "Husserls Manuskripte zu seinem Göttinger Doppelvortrag von 1901", in: K. Schuhmann (ed.), *Husserl Studies* vol. 17, p. 87–123.

Ingarden, Roman (1948), "The Scientific Activity of Kazimierz Twardowski", in: *Studia philosophica* III, p. 17–30.

Kastil, Alfred (2020), "Franz Brentanos Kritik der Antimetaphysiker", in: D. Fisette/G. Fréchette/F. Stadler (eds.), *Franz Brentano and Austrian Philosophy*, Vienna Circle Institute Yearbook, Berlin: Springer, p. 289–308.

Kerry, Benno (1885—1891), "Über Anschauung und ihre psychische Verarbeitung", in: *Vierteljahrsschrift für wissenschaftliche Philosophie* IX, 1885, p. 433–493; X, 1886, p. 419–467; XI, 1887, p. 53–116; XI, 1887, p. 249–307; XIII, 1889, p. 1–124, p. 392–419; XIV, 1890, p. 317–353; XV, 1891, p. 127–167.

Kreibig, Josef Klemens (1905), "Über ein Paradox der Logik Bolzanos", in: *Vierteljahrsschrift für wissenschaftliche Philosophie* vol. 28, p. 375–383.

Kreibig, Josef Klemens (1914), "Bernard Bolzano. Eine Skizze aus der Geschichte der Philosophie in Österreich", in: *Archiv für Geschichte der Philosophie* XXVII, no. 3, p. 273–287.

Künne, Wolfgang (1997), "'Die Ernte wird erscheinen...'. Die Geschichte der Bolzano-Rezeption (1849–1939)", in: H. Ganthaler/O. Neumaier (eds.), *Bolzano und die österreichische Geistesgeschichte*, Sankt Augustin: Academia Verlag, p. 9–82.

Lesky, Erna (1976), *The Vienna Medical School of the 19th Century*, Baltimore: Johns Hopkins University Press.

Maigné, Carole (2017), *Une science autrichienne de la forme: Robert Zimmermann 1824–1898*, Paris: Vrin.
Maigné, Carole (ed.) (2013), *Formalisme esthétique. Prague et Vienne au XIXe siècle*, Paris: Vrin.
Maigné, Carole/Trautmann-Waller, Céline (eds.) (2009), *Formalismes esthétiques et héritage herbartien*, Olms: Hildesheim.
Marty, Anton (1916), *Gesammelte Schriften*, Bd. I, Halle: Niemeyer.
Meister, Richard (1938), "Die Geschichte der Philosophischen Gesellschaft an der Universität Wien 1888–1938", in: R. Reininger (ed.), *50 Jahre Philosophische Gesellschaft an der Universität Wien 1888–1938*, Vienna, p. 3–20.
Misley, Joseph (1818), *Grundriß einer Totalgrundmathesis oder Anwendung der Mathematik auf reingeistige Gegenstände*, Vienna: Gerold.
Moro, Nadia (2009), "De la possibilité et de la nécessité de jugements esthétiques exactement déterminés. Robert Zimmermann interprète de Johann Friedrich Herbart", in: Maigné/Trautmann-Waller (eds.), *Formalismes esthétiques et héritage herbartien*, p. 179–203.
Morscher, Edgar (1997), "Robert Zimmermann – der Vermittler von Bolzanos Gedankengut? Zerstörung einer Legende", in H. Ganthaler/O. Neumaier (eds.), *Bolzano und die österreichische Geistesgeschichte*, Sankt Augustin: Academia Verlag, p. 83–144.
Nawratil, Karl (1969), *Robert Reininger. Leben, Wirken, Persönlichkeit*, Vienna: Verlag der österreichischen Akademie der Wissenschaften.
Nawratil, Karl (1998), "Robert Reininger. Eine Skizze seines Lebens und Wirkens", in T. Binder et al. (eds.), *International Bibliography of Austrian Philosophy*, vol. 8, Amsterdam: Rodopi, p. 1–25.
Payzant, Geoffrey (2002), *Hanslick on the Musically Beautiful: Sixteen Lectures on the Musical Aesthetics of Eduard Hanslick*, Christchurch: Cybereditions.
Reich, Emil (1899), "Robert von Zimmermann. Ein Nachruf", in: *Jahrbuch der Grillparzer-Gesellschaft* 9, p. 322–331.
Ribot, Théodule (1879), *La psychologie allemande contemporaine*, Paris: Germer Baillière.
Rollinger, Robin (1999), *Husserl's Position in the School of Brentano*, Berlin: Springer.
Sauer, Werner (1982), *Österreichische Philosophie zwischen Aufklärung und Restauration*, Amsterdam: Rodopi.
Schlosser, Julius von (2009), "The Vienna School of the History of Art – Review of a Century of Austrian Scholarship in German", in: *Journal of Art Historiography* 1, p. 1–50.
Seiler, Martin (2000), "Empiristische Motive im Denken und Forschen der Wiener Schule der Kunstgeschichte", in: M. Seiler/F. Stadler (eds.), *Kunst, Kunsttheorie und Kunstforschung im wissenschaftlichen Diskurs. In Memoriam Kurt Blaukopf (1914–1999)*, Vienna: ÖBV & HPT, p. 49–86.
Seiler, Martin (2009), "Un 'Manifeste de la philosophie autrichienne'. La nomination du philosophe Robert Zimmermann à l'Université de Vienne (1860–1861)", in: Maigné, C./Trautmann-Waller, C. (eds.), *Formalismes esthétiques et héritage herbartien*, Olms: Hildesheim.
Spitzer, Hugo (1900), "Robert von Zimmermann", in: A. Bettelheim (ed.), *Biographisches Jahrbuch und Deutscher Nekrolog* III, Berlin: Georg Reimer, p. 202–212.
Stachel, Peter (1999), "Leibniz, Bolzano und die Folgen. Zum Denkstil der österreichischen Philosophie, Geistes- und Sozialwissenschaften", in: K. Acham (ed.). *Geschichte der Österreichischen Humanwissenschaften* I, Vienna: Passagen Verlag, p. 253–293.

Stumpf, Carl (1906a), "Zur Einteilung der Wissenschaften", in: *Abhandlungen der Königlich-Preußischen Akademie der Wissenschaften*, philosophisch-historische Classe, Berlin: Verlag der Königlichen Akademie der Wissenschaften, p. 1–94.
Stumpf, Carl (1906b), "Erscheinungen und psychische Funktionen", in: *Abhandlungen der Königlich-Preußischen Akademie der Wissenschaften*, philosophisch-historische Classe, Berlin: Verlag der Königlichen Akademie der Wissenschaften, p. 3–40.
Trautmann-Waller, Céline (2009), "L'héritage herbartien et l'Ecole viennoise de l'histoire de l'art: le cas d'Alois Riegl", in C. Maigné/Trautmann-Waller, C. (eds.), *Formalismes esthétiques et héritage herbartien: Vienne, Prague, Moscou,* Olms: Hildeseim, p. 101–121.
Twardowski, Kazimierz (1894), *Zur Lehre vom Inhalt und Gegenstand der Vorstellungen: Eine psychologische Untersuchung*, Vienna: Hölder; reprinted in Twardowski, *Gesammelte deutsche Werke*, p. 39–122.
Twardowski, Kazimierz (1977), *On the Content and Object of Presentations. A Psychological Investigation*, Eng. trans. R. Grossmann, The Hague: Martinus Nijhoff.
Twardowski, Kazimierz (1999a), *On Actions, Products and other Topics in Philosophy*, J. Brandl/ J. Woleński (eds.), Amsterdam: Rodopi.
Twardowski, Kazimierz (1999b), "Actions and Products. Some Remarks from the Borderline of Psychology, Grammar and Logic", in: Twardowski, *On Actions, Products and other Topics in Philosophy*, p. 103–132.
Twardowski, Kazimierz (1999c), "Self-Portrait", in: Twardowski, *On Actions, Products and other Topics in Philosophy*, p. 17–32.
Twardowski, Kazimierz (2017), *Gesammelte deutsche Werke*, A. Brozek/F. Stadler (eds.), Berlin: Springer.
Twardowski, Kazimierz (2017a), "Über Gebilde und Funktionen", in: Twardowski, *Gesammelte deutsche Werke*, p. 165–191.
Uebel, Thomas (2000), *Vernunftkritik und Wissenschaft: Otto Neurath und der erste Wiener Kreis*, Vienna: Springer.
Varga, Péter András (2015), "Was hat Husserl in Wien außerhalb von Brentanos Philosophie gelernt? Über die Einflüsse auf den frühen Husserl jenseits von Brentano und Bolzano", in: *Husserl Studien* 31, no. 2, p. 95–121.
Wieser, Alfred (1950), *Die Geschichte des Faches Philosophie an der Universität Wien 1848–1938*, Vienna: University of Vienna.
Wiesing, Lambert (2016), *The Visibility of the Image. History and Perspectives of Formal Aesthetics*, London: Bloomsbury.
Winter, Eduard (1933), *Bernard Bolzano und sein Kreis*, Leipzig: Jakob Hegner.
Winter, Eduard (ed.) (1976), *Robert Zimmermanns Philosophische Propädeutik und die Vorlagen aus der Wissenschaftslehre Bernard Bolzanos*, Vienna: Österreichische Akademie der Wissenschaften.
Zimmermann, Robert (1847), *Leibnitz' Monadologie. Deutsch mit einer Abhandlung über Leibnitz' und Herbart's Theorien des wirklichen Geschehens*, Vienna: Braumüller.
Zimmermann, Robert (1848), "Philosophie in Oesterreich und die kaiserliche Akademie der Wissenschaften", in: *Österreichische Blätter für Literatur und Kunst*, p. 315–320.
Zimmermann, Robert (1849a), "Über den wissenschaftlichen Charakter und die philosophische Bedeutung Bernhard Bolzano's", in: *Sitzungsberichte österreichische Akademie der Wissenschaften Philosophisch-historischen Classe*, p. 163–174.
Zimmermann, Robert (1849b), *Leibniz und Herbart. Eine Vergleichung ihrer Monadologien*, Vienna: Braumüller.

Zimmermann, Robert (1850), *Über die jetzige Stellung der Philosophie auf der Universität. Eine Antrittsvorlesung von Rob. Zimmermann, geh. am 15 Apr. 1850*, Olmütz: Hölzel.
Zimmermann, Robert (1852a), *Philosophische Propaedeutic: Prolegomena – Logik – Empirische Psychologie. Zur Einleitung in die Philosophie*, Vienna 1852; 2nd ed. 1860; 3rd ed. 1867.
Zimmermann, Robert (1852b), *Was erwarten wir von der Philosophie? Ein Vortrag beim Antritt des ordentlichen Lehramts der Philosophie an der Prager Hochschule, gehalten am 26. April 1852*, Prague: Credner & Kleinbub.
Zimmermann, Robert (1854), "Die spekulative Ästhetik und die Kritik", in: *Österreichische Blätter für Literatur und Kunst* 6, p. 37–40.
Zimmermann, Robert (1858), *Aesthetik. Erster, historisch-kritischer Teil: Geschichte der Aesthetik als philosophische Wissenschaft*, Vienna: Braumüller.
Zimmermann, Robert (1861), *Philosophie und Erfahrung. Eine Antrittsrede*, Vienna: Braumüller.
Zimmermann, Robert (1862), "Zur Reform der Aesthetik als exakter Wissenschaft", in: *Zeitschrift für exakte Philosophie* 2, p. 309–358; trans. C. Maigné, "Pour une réforme de l'esthétique comme science exacte", in Maigné (ed.) 2013, p. 115–160.
Zimmermann, Robert (1865), *Aesthetik. Zweiter, systematischer Teil: Allgemeine Aesthetik als Formwissenschaft*, Vienna: Braumüller.
Zimmermann, Robert (1871), "Zwei Briefe Herbarts, als Beitrag zu seiner Biographie", in: *Sitzungsberichte der philosophisch-historischen Classe der kaiserlichen Akademie der Wissenschaften* 69, Vienna: Gerold, p. 247–282.
Zimmermann, Robert (1872), "Über Trendelenburg's Einwürfe gegen Herbart's praktische Ideen", in: *Sitzungsberichte der philosophisch-historischen Classe der kaiserlichen Akademie der Wissenschaften* 70, p. 247–282.
Zimmermann, Robert (1873), "Über den Einfluß der Tonlehre auf Herbarts Philosophie", in: *Sitzungsberichte der philosophisch-historischen Classe der kaiserlichen Akademie der Wissenschaften* 73, p. 33–74.
Zimmermann, Robert (1874), "Kant und die positive Philosophie", in: *Sitzungsberichte der philosophisch-historischen Classe der kaiserlichen Akademie der Wissenschaften* 77, p. 31–94.
Zimmermann, Robert (1875), *Schelling's Philosophie der Kunst*, Vienna: Gerold.
Zimmermann, Robert (1876), "Perioden in Herbart's philosophischem Geistesgang. Eine biographische Studie", in: *Sitzungsberichte der philosophisch-historischen Classe der kaiserlichen Akademie der Wissenschaften* 83, Vienna: Gerold, p. 179–234.
Zimmermann, Robert (ed.) (1877), *Briefe von und an Herbart. Aus dessen Nachlaß*, Vienna: Braumüller.
Zimmermann, Robert (1882), *Anthroposophie im Umriss. Entwurf eines Systems idealer Weltansicht auf realistischer Grundlage*, Vienna: Braumüller.
Zimmermann, Robert (1885), "Rezension von: Ed. Hanslick: Vom Musikalisch-Schönen. Siebente vermehrte und verb. Aufl., Leipzig: Barth, 1885", in: *Vierteljahrsschrift für Musikwissenschaft* 1, p. 251–253.
Zimmermann, Robert (1886a), *Über den Antheil Wiens an der deutschen Philosophie*, Vienna: author's edition.
Zimmermann, Robert (1886b), "Kant und Comte in ihrem Verhältnis zur Metaphysik", in: *Sitzungsberichte der philosophisch-historischen Classe der kaiserlichen Akademie der Wissenschaften* 110, p. 3–40.

Zimmermann, Robert (1888), "Wissenschaft und Literatur", in: *Wien 1848–1888*, vol. II, Vienna: Carl Konegen, p. 129–196.
Zimmermann, Robert (1888/1889), "Philosophie und Philosophen in Österreich", in: *Oesterreichisch-Ungarische Revue* 6, p. 177–198, 259–272.
Zimmermann, Robert (1889a), "Die Anfänge der mathematischen Psychologie in Wien", in: *Wiener Zeitung* 37, p. 2–5; 38, p. 2–4; 39, p. 3–5.
Zimmermann, Robert (1893/1894), "Danksagung", in: *Jahresbericht der philosophischen Gesellschaft an der Universität zu Wien*, Vienna: author's edition, p. 5–8.

Riccardo Martinelli

Mental Measurement and the Foundations of Psychology in Nineteenth Century Austrian Philosophy

Abstract: Building upon Kant's concept of intensive magnitude, Herbart developed a method for the measurement of the relative intensity of mental presentations. This relativistic strategy strongly influenced Austrian thinkers and scientists. Among those who further developed this approach, special mention is deserved by Hering, who strongly influences Brentano. With an audacious move, Meinong claimed that both intensive and extensive magnitudes occur within the mental. As a consequence of Russell's criticism of this tenet, Meinong retreated and developed the theory of objects.

1 Kant on Psychology and Intensive Magnitudes

Mental measurement ranks among the most debated philosophical questions of the nineteenth century. Two main groups of solutions can be individuated. In the wake of Herbart, some ventured to measure the relative intensity of certain mental facts, in comparison with the remaining. This "relativistic" approach represented a neat alternative to the "physicalistic" method championed by Fechner, who proposed to measure the physical stimulus and then apply a conversion formula. Needless to say, the choice between these two methods had a tremendous impact on the thorny problem of the foundations of psychology. Herbart's relativistic approach suited the philosophical psychology, in the different forms that flourished especially in Austrian thought; Fechner's physicalistic method preluded to experimental psychology as developed e. g. by Wundt.[1]

In this essay I consider the relativistic approach, drawing attention to five thinkers: Kant, Herbart, Hering, Brentano and Meinong. Given that Meinong is the only Austrian in the list, this choice calls for an explanation. There is little doubt that the names of Herbart and Brentano deserve a place within any history

1 For a comprehensive analysis of Wundt and his relation to Fechner cf. Araujo 2016.

Riccardo Martinelli, Università di Trieste
martinel@units.it

https://doi.org/10.1515/9783110747324-004

of Austrian philosophy.[2] As regards Hering, he taught in Vienna and Prague for almost thirty years, exerting a momentous influence.[3] The most controversial case is that of Kant who, due to his underestimation of formal logic and his diffidence towards empirical psychology, is frequently considered the nemesis of Austrian philosophy. The standard view is that Idealism and Neo-Kantianism ruled undisturbed in Germany, while Realism prevailed among the sober-minded Austrians, who eventually paved the way to analytic philosophy. Though this picture is of course not devoid of wisdom, some caution is recommended especially as far as *Kant himself* (rather than, say, "Idealism" or "Neokantianism") is concerned. Far from ignoring Kant, most Austrian philosophers and scientists – think e. g. of Mach – read him autonomously, occasionally with philological naivety and yet in an original and often philosophically substantial way. Be that as it may, my present task does not include an analysis of Kant's presence within Austrian philosophy. Less ambitiously, I claim that being oblivious to Kant would prevent us to understand the foundations of Herbart's seminal doctrine of the relativity of mental measurement, and then – at least indirectly – of the whole "Austrian way" of measuring mental magnitudes, down to Meinong's doctrine. We will have to admit that Kant's notion of intensive magnitude, introduced in the *Critique of Pure Reason*, is a non-negligible ingredient of the relativistic model of mental measurement.

In the rest of this paragraph I analyze Kant's relevant ideas on the topic. In the next one, I illustrate Herbart's doctrine of mental measurement and contrast it with Fechner's psychophysics. Paragraph 3 is devoted to an exposition of Hering's and Brentano's points of view. The last paragraph deals with Meinong's original ideas on mental measurement and their development.

In the Transcendental Dialectics of *Critique of Pure Reason*, Kant rules out "the notorious question about the community between what thinks and what is extended".[4] This is an obvious corollary of Kant's stance. There being no place for the soul (as substance) among the genuine objects of knowledge, there can be no question concerning the relation of the soul (as substance) with the body. It goes

2 Brentano is even considered by some scholars the champion of Austrian philosophy: cf. Smith 1994. The reasons for the diffusion of Herbartianism in Austria were political rather than philosophical: "Herbartianism […] with its political abstinence and its focus on the formal and empirical aspects of knowledge (both thought to be rather apolitical), seemed an ideal antidote to Kant, who continued to be seen as the 'philosopher of the revolution', and any philosophy that could lead to political unrest." (Landerer/Huemer 2018).

3 After five years at the Josephinum in Vienna, Hering succeeded to Jan Evangelista Purkinje in Prague in 1870, where he would teach until 1895.

4 Kant 1781, p. 438 (A 392); cf. Ameriks 1982.

without saying, then, that none of the traditional solutions for this problem – the physical influence of the empiricists, Leibniz's pre-established harmony, and the supernatural assistance of the occasionalists – retains any validity.[5] Yet for Kant it would be vain to oppose a "dogmatic objection" to their metaphysical claims: the only legitimate counterargument consists of a "well-grounded critical objection".[6] Criticism dissolves the psycho-physical pseudo problem into the task of explaining "[h]ow is outer intuition – namely, that of space (the filling of it by shape and motion) possible at all in a thinking subject?"[7] This question is tackled by Kant in the Transcendental Aesthetics, which overcomes and dismisses any "imaginary science" like Wolff's rational psychology.

In Kant's view, then, the old-fashioned metaphysical doctrine of the soul, i. e. rational psychology, should be replaced by the critique of reason, not by empirical psychology. Empirical psychology, Kant says, "must thus be entirely banned from metaphysics, and is already excluded by the idea of it. [...] It is thus merely a long-accepted foreigner, to whom one grants refuge for a while [...]."[8] Kant adds new details in *Metaphysische Anfangsgründe der Naturwissenschaft*. The "empirical doctrine of the soul" will never deserve the rank of a properly so-called natural science, because

> mathematics is not applicable to the phenomena of inner sense and their laws, the only option one would have would be to take the law of continuity in the flux of inner changes into account – which, however, would be an extension of cognition standing to that which mathematics provides for the doctrine of body approximately as the doctrine of the properties of the straight line stands to the whole of geometry.[9]

In spite of this dismissal of psychology, Kant developed some of the conceptual tools that enabled subsequent thinkers to lay the foundations of empirical psychology. I think of the notion of *intensive magnitude* and of what Kant calls *mathesis intensorum*.[10] In the *Systematic representation of all synthetic principles*, the principles corresponding to the categories of quantity and quality are the *Axioms of intuition* and the *Anticipations of perception*. It is here that Kant distinguishes between extensive and intensive magnitudes. The *Axioms of intuition* rest on the

---

5 Kant 1781, p. 436 (A 390).
6 Kant 1781, p. 437 (A 392).
7 Kant 1781, p. 438 (A 392).
8 Kant 1781, p. 700 (A848/B876–A849/B877).
9 Kant 1786, p. 186. For an analysis of Kant's criticism of empirical psychology cf. Sturm 2009, p. 183 f.
10 Cf. Maier 1968, Martinelli 1996.

principle that "[a]ll intuitions are extensive magnitudes".[11] Extensive magnitudes are defined by Kant as follows:

> I call an extensive magnitude that in which the representation of the parts makes possible the representation of the whole (and therefore necessarily precedes the latter). I cannot represent to myself any line, no matter how small it may be, without drawing it in thought, i.e., successively generating all its parts from one point, and thereby first sketching this intuition.[12]

Phenomena are understood as aggregates of parts: this allows us to apply the "mathematics of extension (geometry)"[13] to the objects of experience, and then to justify *a priori* the geometric axioms.

By contrast, the principle of the *Anticipations of perception* is: "[i]n all appearances the real, which is an object of the sensation, has intensive magnitude, i.e., a degree."[14] In this case, it is no longer a matter of "intuition" (*Anschauung*) and of its application to experience. Kant speaks here of "perception", which necessarily includes a "sensation" (*Empfindung*) caught in its actual development. It is true that sensations decrease in intensity and eventually disappear: but this process has nothing to do with a subtraction of parts. Kant explains:

> Now since sensation in itself is not an objective representation, and in it neither the intuition of space nor that of time is to be encountered, it has, to be sure, no extensive magnitude, but yet it still has a magnitude (and indeed through its apprehension, in which the empirical consciousness can grow in a certain time from nothing = 0 to its given measure), thus it has an intensive magnitude, corresponding to which all objects of perception, insofar as they contain sensation, must be ascribed an intensive magnitude, i.e., a degree of influence on sense.[15]

In other terms, sensations have to do with magnitudes, but not with *extensive* ones. Hence Kant's need for *intensive* magnitudes, defined as follows:

> Now I call that magnitude which can only be apprehended as a unity, and in which multiplicity can only be represented through approximation to negation = 0, intensive magnitude. Thus every reality in the appearance has intensive magnitude, i.e., a degree.[16]

---

11 Kant 1787, p. 286 (B 202). The formulation of the first edition differs slightly.
12 Kant 1787, p. 287 (A 162–163/B 203).
13 Kant 1787, p. 288 (A 163/B 204).
14 Kant 1787, p. 280 (B 207). I discuss the formulation of the first edition below.
15 Kant 1787, p. 290 (B 208).
16 Kant 1787, p. 291 (A 168/B 210).

Perception can be "anticipated", then, by saying that the "reality in the appearance" must possess a certain degree of intensity, different from zero.[17]

Remarkably, the first edition of the *Critique of Pure Reason* contains a rather different version. In 1781 Kant formulated the principle of the *Anticipations of Perception* as follows: "[i]n all appearances *the sensation*, and the real, which corresponds to it in the obiect (*realitas phaenomenon*), has an intensive magnitude, i.e., a degree."[18] The divergence is striking: not just the *reality* which forms the object of sensations (2nd edition 1787), but even *sensations* themselves are intensive magnitudes (1st edition 1781). Clearly, if sensations are magnitudes, they must be measurable. With this admission, unsurprisingly revoked by Kant, the first edition of the *Critique of Pure Reason* implicitly allowed for a measurement of sensations.

It is worth insisting on the importance of this subject for Kant. In the *Prolegomena*, published in 1783, he goes so far as to speak of a "second application of mathematics (*mathesis intensorum*) to natural science".[19] Clearly, the "first" application is based on space and time, given in intuition, and has to do with extensive magnitudes. But this "second" mathematization of nature is not necessarily unimportant. Intensive magnitudes, admitting of a gradual diminution, can be applied to the world of perception, where *reality* – a category, for Kant – manifests itself with its ineliminabile qualitative aspect.

To sum up, Kant leaves no place for rational and empirical psychology in his system (1). Still, dealing with the constitution of physical reality within perception, Kant introduces the concept of intensive magnitude (2) and temporarily goes so far as to admit that sensations have intensive magnitude (3). This ambiguity explains Kant's ubiquitous role within the subsequent controversies on mental measurement and the foundations of psychology. While the enemies of scientific psychology conformed to Kant's rejection of psychology (1), its friends could point at his concessions: Herbart's "relativistic" method of mental measurement relied on Kant's definition of intensive magnitude (2), Fechner's "physicalistic" approach built on Kant's application of intensive magnitude to sensations in the first edition of the *Critique of Pure Reason* (3), to the unspeakable disappointment of the Marburg Neokantians.

---

17 According to this "anticipation", Kant demonstrates e.g. that there is no empty space.

18 Kant 1781, p. 290 (A 166). Emphasis added.

19 Kant 1783, p. 100.

2 Herbart's Psychology

Herbart fiercely opposes any speculative interpretation of criticism: in his view, Kant became an Idealist "against his will".[20] After a Fichtean phase, Herbart becomes dissatisfied with both the Kantian and the Fichtean concept of the Self (*Ich*) – albeit for opposite reasons. In the transcendental deduction of the *Critique of Pure Reason*, the Self is a pure transcendental function, the emptiest and poorest presentation; in Fichte's *Wissenschaftslehre*, by contrast, it turns out to be the unconditional principle underlying the entire philosophical speculation.[21] Steering clear of these opposites, Herbart affirms that we discover the Self in the "inner theater" of consciousness, which reveals a continuous movement and variation: presentations come and go, their intensity sways, they become "stronger or weaker".[22] Herbart likens them to forces endowed with different degrees of intensity. Here we find the magnitudes that make mathematical psychology possible, that is, presentations (*Vorstellungen*).

With this, Kant's intensive magnitudes become the conceptual basis of Herbart's mathematical psychology. A confirmation comes from Herbart's reaction to Kant's criticism of Moses Mendelssohn, who claimed that the soul can never cease to exist, for metaphysical reasons. For a simple substance, Mendelssohn argued, a leap into Nothingness is unthinkable. Only God could decree its annihilation, but he never would: therefore, the soul is necessarily immortal.[23] In the second edition of the *Critique of Pure Reason*, Kant replies that nothing prevents the soul from undergoing a process of progressive "remission", eventually culminating in its complete extinction. For Kant, Mendelssohn

> did not consider that even if we allow the soul this simple nature, namely, that it contains no manifold [of parts] *outside one another*, and hence no extensive magnitude, one nevertheless cannot deny to it, any more than to any other existence, an intensive magnitude, i.e., a degree of reality in regard to all its faculties, indeed to everything in general that constitutes its existence, which might diminish through all the infinitely many smaller degrees; and thus the supposed substance [...] could be transformed into nothing, although

20 Herbart 1824, p. 179. Herbart's non-Kantian reputation "is strikingly at odds with his own self-conception" (Beiser 2015, p. 89). Beiser argues that psychology is exactly where Herbart diverges from Kant. Notwithstanding this undeniable fact, Herbart builds his psychology upon Kant's concept of intensive magnitude (see below). On the Kantian origin of the conceptual tools of Herbart's psychology, cf. Leary 1980.

21 Herbart 1822, p. 108; cf. also Herbart 1824, p. 222ff., 241.

22 Herbart 1824, p. 196, 251.

23 Mendelssohn 1767, p. 71. Mendelssohn's argument has been simplified here; for a more comprehensive analysis, cf. Martinelli 2002.

> not by disintegration, but by a gradual remission (*remissio*) of all its powers (hence, if I may be allowed to use this expression, through elanguescence).[24]

Herbart disagrees: that one cannot deny to the soul "any more than to any other existence, an intensive magnitude", is wrong. In Herbart's view, in fact, *real beings* (*Wesen*) have no intensive magnitude. By contrast, as far as *presentations* are concerned, intensive magnitudes play a crucial role. Herbart clearly states that a diminution and vanishing of *presentations* is not only "thinkable", but indeed "necessary to the explanation of the Self".[25] Herbart's presentations are intensive magnitudes in Kant's sense.

More precisely, Herbart's psychology is based on the measurement of two mental magnitudes: the "strength" and "the degree of inhibition" (*Hemmung*) between two given presentations.[26] Inhibition is one of the most important technical terms of Herbart's psychology: striving for preservation, presentations conflict with each other: they try to prevail and "inhibit" the others.[27] The dominating presentation pushes all the others away, forcing them to decrease in intensity and to sink below the threshold (*Schwelle*).[28] Herbart writes:

> If we now connect [...] the difference between representations in terms of their intensity [*Stärke*], on the one hand, with the magnitude of their mutual opposition, on the other, then this yields in each case the magnitude of the resulting dimming, inhibition, [and] striving; and so too [the magnitude] of the remaining actual representation. Here calculation [*Rechnung*] finds its appropriate matter.[29]

The consequences of Herbart's move are remarkable. Kant considered *mathesis intensorum* a "second application" of mathematics to the science of nature within perception, the first application being that obtained from the pure intuitions of space and time. By contrast, Herbart elevates the *mathesis intensorum* to the rank of a *first application* of mathematics to psychology, while the alleged "pure

---

24 Kant 1787, p. 449 (B 414).

25 Herbart 1824, p. 270.

26 Herbart 1822, p. 102.

27 "In the process of striving against each other, some presentations are driven out of consciousness, which reduces the overall conflicting power and diminishes the inhibition sum. According to Herbart, this process can be expressed in the following formula: $t = \log \cdot S/S-\sigma$, with S being the inhibition sum and σ its already sunken part which can be determined as $\sigma = S(1- e-t)$." (Landerer/Huemer 2018). This explanation exemplifies the technicality of Herbart's psychology.

28 Herbart 1824, p. 275.

29 Herbart 1824, p. 279. English translation quoted from Kim 2015.

intuitions" of space and time become merely residual consequences of the interaction between different series of presentations within the mind.[30]

The publication of Fechner's *Elemente der Psychophysik* in 1860 is a milestone in the development of this debate. Modern readers may fail to appreciate the outstanding importance of Fechner's book and its heavy impact upon generations of researchers. Starting from a monistic world view, Fechner affirms that mental and physical magnitudes are nothing but the two sides of the same ontological coin. He famously writes: "Leibniz says: [...] two clocks mounted on the same board adjust their movement to each other by means of their common attachment [...]; this is the usual dualistic notion of the mind-body relation." However, "Leibniz has left out one point of view – the most simple possible. They can keep time harmoniously – indeed never differ – because they are not really two different clocks."[31] Accordingly, Fechner tries to circumvent the technical problem of mental measurement by measuring the stimuli, i. e., the physical entities regularly conjoined to sensations. Building upon Ernst Heinrich Weber's experiments, Fechner finds a mathematical correlation between the intensity of the stimulus and the intensity of the corresponding sensation: the celebrated logarithmic formula $E = k\ log(R)$, that is, the intensity of a sensation (E for *Empfindung*) corresponds to the logarithm of the intensity of the stimulus (R for *Reiz*).[32]

In this essay I will not deal with Fechner's psychophysics. Suffice to say once again that Herbart and Fechner defined two deeply diverging solutions for the problem of mental measurement: Fechner's measurement of physical stimuli contrasts with Herbart's insistence on the relative value of any mental measurement.

3 From Hering to Brentano

Hering begins his pervasive attack on psychophysics of 1874 by recalling his teachers Ernst H. Weber and Gustav T. Fechner with sincere deference. He affirms that his conception of the relationship between body and soul "stands in better agreement with the philosophy of Fechner than his own psychophysical law".[33]

---

30 Herbart 1824, p. 427ff.

31 Fechner 1860, p. 4.

32 For an excellent survey of Fechner's psychophysics, cf. Woodward 2018. A comprehensive account is Heidelberger 1993.

33 Hering 1875, p. 327f. Hering later called himself "an enthusiastic follower although not a personal disciple" of Johannes Müller (Hering 1884, p. 28).

In fact, Hering assumed that the sensory and motor processes found a common origin in the phylogenetic evolution, and in a famous speech of 1870, he went so far as to consider memory a common trait between the two realms.[34] This said, Hering was far from Fechner's idealistic monism and rather insisted on empirical research. In his view, even Fechner's unified formula for all sensory fields is *a priori* implausible and can be empirically disproved. The impressions of weight are proportional to the stimuli; by contrast, a quasi-logarithmic relation holds for visual sensations: Fechner's alleged universal law can be validated only for the intensity of sounds.[35] To be sure, Hering does agree that a general formula should be searched for. But this formula has nothing to do with the relation between sensations and physical stimuli. Rather, there is proportionality between sensations and the corresponding *physiological* excitement: "[t]he magnitude or intensity of the mental process increases and decreases proportionally with the physical processes that directly cause them."[36] Hering rules out Fechner's model of mental measurement from the very beginning.[37]

Against his scientific archenemy Hermann Helmholtz, Hering insists that the investigation of stimuli and of the peripheral nervous system is not enough, because

> [a]ll living substance, especially nerve-matter, has the peculiarity that every stimulation produced in a limited region at once spreads to the adjoining parts. It continues spreading as long as it meets with any substance which is capable of being similarly stimulated and which, so to speak, responds to such stimulation.[38]

This happens, most notably, in the central nervous system which, in spite of "its being a compound of many thousands of cells and fibers", "forms one coherent entirety" and "is in communication with all organs".[39] As a consequence, the same objective stimulus may evoke different perceptual results: with this, Hering

34 Hering 1870; cf. Turner 1994, p. 57.
35 Hering 1875, p. 346.
36 Hering 1874, p. 330.
37 "Hering's general psychophysical law is about the relation between *physiological excitement and sensations*, and cannot be expressed by a logarithmic function; his law, by contrast, has nothing to do with the relation *between physical stimuli and physiological excitement*: nothing at all, in fact, can be said about that." (Hillebrand 1918, p. 18 f.). Franz Hillebrand was a student of Brentano and Anton Marty, who then attended Hering's lectures in Prague.
38 Hering 1870, p. 37.
39 Hering 1870, p. 15f.

anticipates the subsequent phenomenological or holistic takes on psychology.[40] For instance, if an object casts a shadow on a portion of a white sheet, we call the shaded part "darker" and not "grey", although "the reflected light may have exactly the same intensity and composition" of the light reflected from a gray sheet of paper under full light.[41] To explain this fact, Hering hints at the general characteristics of the corresponding brain area, in which the "optical experience of a whole life" is preserved and organized. In fact, the "nervous substance faithfully preserves the records of process often performed".[42] With a suggestively musical and quintessentially anti-Helmholtzian image, Hering hints at the "resonance of our entire sensory apparatus".[43]

As far as mental measurement is concerned, Hering develops a highly interesting method based upon the sole internal relations between sensations, independently of the stimuli. Invading so to speak Fechner's field with an unmistakably Herbartian methodology, Hering develops a "relativistic" method for measuring *sensations*. For instance, his quantification of visual sensations begins with the "neutral gray", i. e. the perfect intermediate between black and white. Provided that the neutral grey (N) contains black (B) and white (W) in equal proportions (B = W), Hering assigns it the numerical value 1: "B / W (or W / B) = 1". Then Hering considers the chromatic shades situated halfway between the intermediate gray (N) and the two extremes (B or W), i. e. at 1/4 and 3/4 of the scale, etc.[44] Disregarding further technicalities, it is clear that Hering avoids any reference to the stimuli and rather aims at the relative value of mental measurements in the case of visual sensations. As a consequence, Hering criticizes the standard view that black and white correspond to a minimum and a maximum of intensity of the visual sense. In his view, both white and black are positive sensations, not

---

40 "I am convinced that Ewald Hering has received inadequate treatment [...] as to his great influence upon experimental psychology in general and especially upon those in Brentano's phenomenological camp. I regret more than any other sin which my critics have mentioned this injustice to his memory." (Boring 1950, p. xii). Against the "stereotype of Hering as phenomenologist", Hurvich stresses that "Hering was an experimentalist of a special breed": Hurvich 1969, p. 500. This means that Hering's writings match all the criteria of scientific research, against the allegation of *naive* phenomenalism. Surprisingly, there is no monograph devoted to Hering so far.

41 Hering 1874, p. 66 f. [101]; cf. Casati 1994.

42 The unconscious "sensory memory" is physiological: "[t]o a physical consideration [...] unconscious and material mean the same" (Hering 1870, p. 10).

43 Hering 1874, p. 68 [103]. Helmholtz had developed for the sensations of sound the so-called "one fiber/one sensation" doctrine (cf. Hatfield 1994), sharply attacked by Hering.

44 Hering 1874, p. 59.

degrees of intensity: "as a mere sensation, black is just as positive as white".[45] In support of this stance, Hering experimentally demonstrates that black is not the default sensation of a resting eye, but arises from simultaneous or subsequent contrast with high illumination. The historical importance of this argument is remarkable, because it influenced Brentano's elaboration of his mature doctrine of intensity.

In *Psychology from an Empirical Standpoint* of 1874, Brentano insists that psychology makes no exception to the scientific method. Its first step is the induction of general laws, followed by the deduction of particular laws and their experimental verification. However, Brentano admits, the case in point requires some caveats, especially with regard to the former moment (induction). In fact, there are two limitations: 1) mental facts depend on concomitant physiological conditions, and 2) there is no way to measure their intensity.

> These, then, are the two factors which prevent us from acquiring an accurate conception of the highest laws of mental succession: first, they are only empirical laws dependent upon the variable influences of unexplored physiological processes; secondly, the intensity of mental phenomena, which is really one of the decisive factors, cannot as yet be subjected to exact measurement.[46]

Despite the latter admission, Brentano keeps a surprisingly open mind when it comes to the possibility of mental measurement. To be sure, his main purpose is *not* the development of a quantitative psychology based on mental measurements. Still, he concedes its principal legitimacy, allowing for an epistemic model that (at least, in 1874) goes hand in hand with his descriptive approach to psychology.[47] Let us illustrate some aspects of this model.

In his criticism of psychophysics, Brentano keeps far from the destructive attitude, say, of the neo-Kantians, who followed the second edition of Kant's *Critique of Pure Reason*, denying that sensations can ever admit of intensity.[48] Less radically, Brentano attempts at *reforming* psychophysics.[49] After a thorough

---

45 Hering 1874, p. 54.

46 Brentano 1874, p. 53.

47 "Brentano's critique of Fechner is theoretically important because it highlights a series of essential steps in his elaboration of *descriptive psychology* as the development of a *sui generis* inner psychophysics." (Albertazzi 2006, p. 103). In my view, both of these models – descriptive psychology and "*sui generis* inner psychophysics", still coexist in 1874. Herbartian relativism concerning mental measurement occupies a relevant place in that "*sui generis*".

48 Most remarkably, Cohen 1883; cf. Martinelli 1999, p. 73.

49 "The dispute with Fechner is of strategic importance for Brentano's theory of sensation", both in the early version and in the late one (Seron 2011, p. 87). The correspondence between

discussion of Fechner's law, he concludes that "if the strengths of the physical stimuli increase by the same number of times, the intensities of the sensations also increase by the same number of times".[50] In his view, then, proportionality takes the place of the logarithmic function. Of course, this is a substantial emendation, that ends by undermining Fechner's construct and calls for a different stance in psychology.[51] However, it ought to be noted that Brentano denies neither that sensations have an intensity, nor that this intensity can be measured.

Brentano also rejects Herbart's way of applying mathematics to psychology. He writes:

> It was Herbart who first emphasized the necessity of such measurements. The merit which he earned thereby is just as generally recognized as is the complete failure of his attempt to discover actual determinations of quantity. The arbitrariness of the ultimate principle upon which he bases his mathematical psychology cannot be compensated for by his consistent adherence to the rigorous laws of mathematics in deducing consequences.[52]

Rather than the abandonment of mental measurement altogether, the consequence of this criticism for Brentano is the development of a new, simplified method, which is principally in line with Herbart's relativistic approach, in the above defined sense. In fact, Brentano concedes that every mental presentation principally has an intensity, and that it is measurable. This holds for mental acts of any kind, including judgments and phenomena of love and hate (which have, in addition, a specific intensity of their own).

No wonder, then, that Brentano frequently refers to mental measurement in his *Psychology*. This happens, for instance, when it comes to the non-existence of

---

Brentano and Fechner has been published (Brentano/Fechner 2015) with an exhaustive introduction (Antonelli 2015).

50 Brentano 1874, p. 51; Brentano's objection is complicated by a hair-splitting distinction between "equal" and "equally noticeable" (*gleichmerklichen*) increments, which Fechner found "incomprehensible" (*unverständlich*) (Brentano/Fechner 2015, p. 120).

51 Brentano complains: "[...] the possibility of measuring intensities according to their method is restricted entirely to those phenomena which are produced by external stimulation of the sense organs. We still lack, therefore, a measure of intensity for all mental phenomena which have their foundation in physical processes within the organism or which are caused by other mental phenomena. But the majority of mental phenomena including the most important ones belong in this category: the whole class of desires and actions of the will, as well as convictions and opinions of all kinds, and a wide range of presentations which have their origin in the imagination. Of all mental phenomena, sensations alone, and not even all of them, remain measurable." (Brentano 1874, p. 52).

52 Brentano 1874, p. 50.

unconscious mental phenomena, a basic principle of his doctrine. In the context of this demonstration, Brentano states that

> [t]he intensity of the act of presentation is always equal to the intensity with which the object that is presented appears to us; in other words, it is equal to the intensity of the phenomenon which constitutes the content of the presentation.[53]

Integrating his explanation, we can formulate this principle in symbols:

$$i\,(R) = k\,i\,(O)$$

with i = intensity, R = presentation, and O = intentional object. This formula allows us to summarize Brentano's early view. In the case of sensations, the intensity of the mental act is *proportional* to that of the object: *k* is a positive number ($k>0$). In the case of non-sensory presentations, there is a stronger condition: the intensity of the mental act is *the same* ($k=1$). A further complication must be mentioned at least in passing: for Brentano, the intensity of the act of inner consciousness, by means of which a certain act, say, of presentation, is known to us, obeys the same law: "[...] if the intensity of the presentation is always equal to the intensity of the phenomenon which constitutes its content, it is clear that the intensity of *the presentation of a presentation* must also be equal to the intensity with which this latter presentation manifests itself."[54] In addition, this act of inner consciousness (technically speaking: a judgement), has an intensity of its own, i. e. the degree of our belief in the existence of its object. Brentano concludes:

> But there can be no doubt in this regard. Both intensities must be equal, if inner perception is indeed infallible. Just as inner perception cannot confuse seeing and hearing, neither can it mistake a strong auditory sensation for a faint one nor a faint for a strong one. So we come to the conclusion that the intensity of the presentation of every conscious presentation is equal to the intensity of that presentation.[55]

In 1874, then, Brentano does not shy away from expressing the pivotal tenet of his psychology – the evidence of inner perception – in terms of measurable intensities.

In the 1880's Brentano completely changes his mind on this whole question. He now draws a distinction between those mental phenomena that have intensity (*sensations*), and those that are totally devoid of it (all the remaining). Brentano urges that certain presentations, e. g. that of a certain number, have no intensity;

53 Brentano 1874, p. 93.
54 Brentano 1874, p. 93, emphasis added.
55 Brentano 1874, p. 93.

similarly, the different degrees of belief (e. g. in the existence of a horse, or a centaur) have nothing to do with intensity. In 1874, Brentano admits, "I followed the received opinion, according to which degrees of conviction were to be understood as differences in intensity. But this opinion, as I have now seen, is a mistaken one." Accordingly, also

> the degree of preference and the will's degree of determination are not at all comparable to the degrees of intensity of a sensation. And [...] the view that every mental reference exhibits intensity in the strict sense must be given up since we even find presentations (as for example that of the number "three" in general) without intensity.[56]

These words are taken form the sixth Appendix of his book *Von der Klassifikation der psychischen Phänomenen* of 1911, entitled "On the Impossibility of Ascribing Intensity to Every Mental Reference and in Particular the Impossibility of Understanding Degrees of Conviction and Preference as Differences of Intensity". This short annotation at the end of the book may leave the reader with the impression of a minor adjustment within a basically unaffected body of doctrine. In fact, Brentano's new stance amounts to a remarkable revision: the sections of *Psychology from an Empirical Standpoint* where he refers to the intensity of presentations, judgements and emotions (see e. g. the quotations above) simply become untenable.

His adherence to Helmholtz's doctrine of the "specific energies" of the sensory apparatus counts among the strongest reasons of Brentano's change of mind.[57] Interestingly, Brentano's new stance reveals the parallel influence of Hering's tenet that intensity cannot be considered a general characteristic of all sensations. With his refusal to ascribe intensity to visual sensations, Brentano argues, Hering unmasked the current dogmas on intensity. Since Hering "went too far" with that, the necessary correction is provided by Brentano's new doctrine of sensations.[58] Intensity becomes now a spatial (or *quasi-spatial*) notion: Brentano thinks of an internal "space of sensation" (*Empfindungsraum*).[59] When every spot of this "intentional" space is filled by sensory qualities, intensity reaches a maximum; when some of the spots are empty, intensity sinks.[60] Basically, this is why intensity applies only to sensations and not to intellectual presentations.

---

56 Brentano 1911, p. 223.

57 Cf. Martinelli 2017, p. 344.

58 Brentano 1896, p. 84; cf. also p. 74, 76.

59 Brentano 1907, p. 70.

60 This mechanism, Brentano believes, also accounts for mixed sensory qualities. If the spots are filled with different qualities (e. g. red and blue), the overall impression is that of an

In sum, after an early phase in which he endorses a revised version of the mainstream doctrines of mental measurement, Brentano eventually denies that intensity is applicable in the mental sphere. Intensity applies now to sensations only, under exclusion of other mental phenomena. Even in the case of sensations, intensity ceases to be a phenomenal trait: rather, it is a quasi-spatial characteristic of the intentional object of sensory presentations. Mental measurement has now become an irrelevant task for Brentano.

4 Meinong and beyond

Meinong devotes much attention to Weber's law, intensive magnitudes, and the problem of mental measurement. In a dense essay of 1896 entitled *Über die Bedeutung des Weber'schen Gesetzes. Beiträge zur Psychologie des Vergleichens und des Messens,* Meinong targets Johannes von Kries, who had argued that intensive magnitudes must be invariably reduced to the three dimensions of space, time and mass. Since this reduction is manifestly impossible in the case of sensations, Kries concluded, Fechner's psychophysics is pointless.[61] Meinong objects that this criticism depends on the unjustified presupposition that "the physical" encompasses both extensive (e. g. length) and intensive (e. g. speed) magnitudes, while "the mental" encompasses exclusively intensive magnitudes. Contrasting this view, Meinong claims that *both* kinds of magnitudes occur in the physical *and* in the mental world as well.[62] To say the least, this stance is unconventional. Herbart's widespread view that the mental is characterized exclusively by *intensive* magnitudes is revoked: but Meinong does not conclude (like Kries and many others) that the mental encompasses *no measurable magnitudes at all:* on the contrary, he assumes that there is *extension* too within the mind. But how to allow – challenging Descartes' dichotomy – for "extensiveness" in the mental world? Meinong's solution rests on the distinction between act and content: "sensing" as such (*das Empfinden*) is exclusively intensive, yet "what is sensed" (*das Empfundene*) is not subjected to this restriction – and it is mental, in turn.[63] From the

intermediary quality between them (e. g. purple). Almost entirely devoid of empirical support, this metaphysics of sensibility failed to convince even Brentano's pupils.

61 Kries 1882. "Hering and Helmholtz had exchanged polemics directly in the 1860s, and thereafter Hering and his students warred with scientists they regarded as Helmholtz's disciples and surrogates, notably Adolf Fick, Arthur König, and Johannes von Kries." (Turner 1994, p. 5; cf. also p. 181, 190).

62 Meinong 1896, p. 322.

63 Meinong 1896, p. 323.

point of view of its theoretical measurability, a "represented segment" behaves like any truly existing segment: the difficulties about its measurement are merely practical.

The core of Meinong's doctrine lies in his definition of the concept of magnitude. What is typical of all magnitudes is that they "limit against zero".[64] A good explanation of this stance has been given by Bertrand Russell:

> Herr Meinong [...] regards zero as the contradictory opposite of each magnitude of its kind [...]. It can hardly be regarded as true to say that a pain is a zero pleasure. On the other hand, a zero pleasure is said to be no pleasure, and this is evidently what Herr Meinong means.[65]

Now, we can always think of a certain degree between a given magnitude and zero – say, between a state of pleasure and zero pleasure. This leads Meinong to a further definition: "[m]agnitude is [...] that which allows the interpolation of terms between itself and its contradictory opposite", i. e. zero.[66]

This definition of *magnitude* is strikingly similar to the definition of *intensive magnitude* given by Kant, duly referred to by Meinong in a footnote.[67] What for Kant was the thorny, special case of intensive reality, becomes for Meinong the basic concept of magnitude as such. More specifically, Meinong allows for different classifications, e. g. between *divisible* and *non-divisible* magnitudes. In the former case, measurement is a "comparison of parts" (*Teilvergleichung*);[68] in the latter, Meinong resorts to what he calls "substitutive" (*surrogativ*) measurement. A certain degree of temperature, for example, is non-divisible: it is obviously not made up of multiple "partial" temperatures. Therefore, it is necessary to refer to a substitute, i. e. the quicksilver of the thermometer.[69] It is characteristic of Meinong's view that this conceptual apparatus can be indifferently applied to

64 "Es ist allen Größen charakteristisch, gegen Null zu limitieren." (Meinong 1896, p. 218).

65 Russell 1903, p. 184f.

66 Meinong 1896, p. 218. This definition was translated into English by Russell 1903, p. 168f.; cf. Guigon 2005, p. 259.

67 Meinong 1896, p. 218, fn. 7. "Meinong's starting point is clearly Kant's definition of an intensive magnitude." (Guigon 2005, p. 257). Tegtmeier claims – wrongly, in my view – that the Kantian root of Meinong's stance on measurement is merely accessory; Tegtmeier 1996, p. 162. Tegtmeier contrasts Meinong's *realist* view that "quantities exist independently of our measurment and that by measuring we only apprehend quantities and relations between them", with the *positivist* one, endorsed e. g. by Mach. "I am convinced – he writes – that the positivist theories of measurement are inadequate, that Meinong's criticism of these is still sound, that his own theory is more satisfactory and deserves renewed interest." (Tegtmeier 1996, p. 161).

68 Meinong 1896, p. 271.

69 Meinong 1896, p. 277.

physical or mental magnitudes. His conclusion on mental measurement is: “no authentic mental measurement is immediate; and there is no substitutive mental measurement, in which the substitute is liable to immediate measurement”.[70]

The historical context of these ideas is quite interesting. A sharp criticism by Bertrand Russell shows that Meinong comes very close here to Francis Herbert Bradley, who revived Hegel’s theses on the dialectic convertibility of intensive and extensive quantities.[71] In his review of Meinong’s “excellent” essay, Russell notes that Meinong “does not accept the view that psychical quantities must be intensive. He urges, in agreement with Mr. Bradley, that psychical quantities may be extensive, since the presented, as such, is psychical.”[72] This has undesirable consequences.

> If the presented, as such, is psychical, then every possible object of experience is psychical. This leads either to the philosophy of Berkeley, or to an unknowable thing in itself. To urge, as Herr Meinong does, that imagined space is measurable and divisible, though purely psychical, seems either irrelevant or untrue. For imagined space is as little mental as real space; it differs from real space only in the fact that it does not exist: while the imagination of space, which does exist, is not divisible.[73]

Meinong’s reaction is of utmost importance. In his essay *On Objects of Higher Order* of 1899, he avows that Russell’s criticism is “completely justified”.[74] It is precisely from this criticism that Meinong proceeds to ascribe *pseudo*-existence to immanent objects, and then to develop his theory of objects.[75] Meinong’s pupil Ernst Mally will take care of adjusting the doctrine of measurement to the new conceptual framework of the *Gegenstandstheorie*.[76]

By all divergencies, Russell remarkably agrees with Meinong on measurement.[77] In *Principles of Mathematics* (1903) Russell defines the concept of magni-

---

70 Meinong 1896, p. 329.

71 Bradley 1895a, 1895b. Following Hegel’s *Logic*, Bradley thinks of quantity in terms of a modification that occurs within quality.

72 Russell 1899; the reference is to Bradley 1895a.

73 Russell 1899.

74 Meinong 1899, p. 383, fn. 1.

75 In turn, the theory of object will be famously criticized by the implacable Russell in “On Denoting” (Russell 1905).

76 Mally 1904. As regards the presence of theory of objects already in Meinong 1896, cf. Potrc/Vospernik 1996, p. 191.

77 Guigon observes that Meinong 1896 “is the only work of Meinong ever substantially endorsed by Russell. Part III and half of Part IV of *The Principles* [*of Mathematics*] are explicitly grounded on Meinong’s work. If *The Principles* is an important work in the history of Analytic Philosophy,

tude as follows: "anything which is greater or less than something else".[78] Russell comments:

> Thus it would seem that Herr Meinong's theory [...] is substantially correct; it requires emendation [...] only in this, that a zero magnitude is the denial of the defining concept of a kind of magnitudes, not the denial of any one particular magnitude, or of all of them.[79]

With this, the Austrian legacy on the problem of mental measurement takes an unexpected turn. Bypassing Herbart and Brentano, Meinong resorts to Kant, whose definition of intensive magnitude he adopts and generalizes. With this same move, curiously enough, Meinong shakes hands with Bradley's Hegelianism and lays the groundwork for Russell's logicist doctrine of measurement.

References

Albertazzi, Liliana (2006), *Immanent Realism. An Introduction to Brentano*, Dordrecht: Springer.

Ameriks, Karl (1982), *Kant's Theory of Mind. An Analysis of the Paralogisms of Pure Reason*, Oxford: Clarendon.

Antonelli, Mauro (2015), "Franz Brentano und Gustav Theodor Fechner über Psychophysik", in: Franz Brentano/Gustav Theodor Fechner, *Briefwechsel über Psychophysik, 1874–1878*, ed. by M. Antonelli, Berlin: De Gruyter, p. 3–75.

Araujo de Freitas, Saulo (2016), *Wundt and the Philosophical Foundations of Psychology. A Reappraisal*, Cham: Springer.

Boring, Edwin G. (1950), *A History of Experimental Psychology*, 2nd ed., New York: Appleton-Century-Crofts.

Bradley, Francis Herbert (1895a), "What Do We Mean by the Intensity of Psychical States?", in: *Mind* 4 (13) (n.s.), p. 1–27.

Bradley, Francis Herbert (1895b), "In What Sense Are Psychical States Extended?", in: *Mind* 4 (13) (n.s.), p. 225–235.

Brentano, Franz (1874), *Psychologie vom empirischen Standpunkte*, Leipzig: Duncker & Humblot (quoted from the English translation: Brentano 2009).

Brentano, Franz (1896), "Über Individuation, multiple Qualität und Intensität sinnlicher Erscheinungen", in: *Untersuchungen zur Sinnespsychologie*, 2nd ed., ed. by R. Chisholm and R. Fabian, Hamburg: Meiner 1979, p. 66–89.

Brentano, Franz (1911), "Von der Unmöglichkeit, jeder psychischen Beziehung eine Intensität zuzuerkennen und insbesondere die Grade der Überzeugung und Bevorzugung als Unterschiede der Intensität zu fassen", in: *Von der Klassifikation der psychischen Phänomene*.

---

which of course it is, Meinong's work on Weber's Law is a non-negligible part of this history." (Guigon 2005, p. 256).

78 Russell 1903, p. 159.

79 Russell 1903, p. 187.

Neue, durch Nachträge stark vermehrte Ausgabe der betreffenden Kapitel der Psychologie vom empirischen Standpunkt, Duncker & Humblot: Leipzig (quoted from the English translation: Brentano 2009, p. 223–224).
Brentano, Franz (2009), *Psychology from an Empirical Standpoint*, ed. by L. McAlister, London/New York: Routledge.
Brentano, Franz/Fechner, Gustav Theodor (2015), *Briefwechsel über Psychophysik, 1874–1878*, ed. by M. Antonelli, Berlin: De Gruyter.
Bridgeman, Bruce/Stark, Lawrence (1977), "Introduction", in: Ewald Hering, *The Theory of Binocular Vision*, New York/London: Plenum Press, p. 1–13.
Casati, Roberto (1994), "The Concept of Sehding from Hering to Katz", in: *Gestalt Psychology: Its Origins, Foundations, and Influence*, ed. by S. Poggi, Florence: Olschki, p. 21–57.
Cohen, Hermann (1883), *Das Prinzip der Infinitesimal-Methode und seine Geschichte. Ein Kapitel zur Grundlegung der Erkenntniskritik*, in: *Werke*, vol. 5, ed. by the Hermann Cohen-Archiv, Hildesheim: Olms 1984.
Fechner, Gustav Theodor (1860), *Elemente der Psychophysik*, 2 vols., Leipzig: Breitkopf & Härtel.
Guigon, Ghislain (2005), "Meinong on Magnitudes and Measurement", in: *Meinongian Issues in Contemporary Italian Philosophy*, ed. by A. Schramm, Frankfurt a. M.: Ontos, p. 255–296.
Hatfield, Gary (1994), "Helmholtz and Classicism. The Science of Aesthetics and the Aesthetics of Science", in: *Hermann von Helmholtz and the Foundation of Nineteenth-Century Science*, ed. by David Cahan, Berkeley: University of California Press, p. 522–558.
Heidelberger, Michael (1993), *Die innere Seite der Natur. Gustav Theodor Fechners wissenschaftliche-philosophische Weltauffassung*, Frankfurt a. M.: Klostermann.
Helmholtz, Hermann von (1887), "Zählen und Messen, erkenntnistheoretisch betrachtet", in: *Philosophische Aufsätze, Eduard Zeller zu seinem fünfzigjährigen Doctor-Jubiläum gewidmet*, Leipzig: Fues, p. 17–52.
Herbart, Johann Friedrich, (1822), "Über Möglichkeit und Nothwendigkeit, Mathematik auf Psychologie anzuwenden", in: *Sämtliche Werke*, ed. by K. Kehrbach and O. Flügel, Langensalza, 1887–1915; repr. Aalen: Scientia 1989, vol. 5, p. 91–122.
Herbart, Johann Friedrich (1824), *Psychologie als Wissenschaft, neu gegründet auf Erfahrung, Metaphysik und Mathematik*. Erster, synthetischer Teil, in: *Sämtliche Werke*, ed. by K. Kehrbach and O. Flügel, Langensalza, 1887–1915; repr. Aalen: Scientia 1989, vol. 5, p. 177–402.
Herbart, Johann Friedrich (1825), *Psychologie als Wissenschaft, neu gegründet auf Erfahrung, Metaphysik und Mathematik*. Zweiter, analytischer Teil, in: *Sämtliche Werke*, ed. by K. Kehrbach and O. Flügel, Langensalza, 1887–1915; repr. Aalen: Scientia 1989, vol. 6, p. 1–338.
Hering, Ewald (1870), "Über das Gedächtnis als eine allgemeine Funktion der organisierten Materie. Vortrag gehalten in der feierlichen Sitzung der Kaiserlichen Akademie der Wissenschaften in Wien am 30. Mai 1870" (quoted from the English translation: *Memory. Lectures on the Specific Energies of the Nervous System*, Chicago: Open Court 1913).
Hering, Ewald (1874), "Zur Lehre vom Lichtsinne. IV: Über die sogenannte Intensität der Lichtempfindung und über die Empfindung des Schwarzen", in: *Sitzungsberichte der Kaiserlichen Akademie der Wissenschaften. Mathematisch-naturwissenschaftliche Klasse*, Vienna, p. 85–104.
Hering, Ewald (1875), "Zur Lehre von der Beziehung zwischen Leib und Seele. I: Über Fechners psychophysisches Gesetz", in: *Sitzungsberichte der Kaiserlichen Akademie der Wissenschaften. Mathematisch-naturwissenschaftliche Klasse*, Vienna, p. 310–348.

Hering, Ewald (1884), "Über die spezifischen Energien des Nervensystems", in: *Lotos* 5 (quoted from the English translation in: *Memory. Lectures on the Specific Energies of the Nervous System*, Chicago: Open Court 1913).

Hillebrand, Franz (1919), *Ewald Hering. Ein Gedenkwort der Psychophysik*, Leipzig: Springer.

Hurvich, Leo M. (1969), "Hering and the Scientific Establishment", in: *American Psychologist* 24, p. 497–514.

Kant, Immanuel (1781) *Kritik der reinen Vernunft* (A), in *Kants Werke. Akademie-Ausgabe*, vol. 4 (quoted from the English translation: Kant 1998).

Kant, Immanuel (1783), *Prolegomena zu einer jeden künftigen Metaphysik die als Wissenschaft wird auftreten können*, in *Kants Werke, Akademie-Ausgabe*, vol. 4.

Kant, Immanuel (1786) *Metaphysische Anfangsgründe der Naturwissenschaft*, in *Kants Werke. Akademie-Ausgabe*, vol. 4.

Kant, Immanuel (1787) *Kritik der reinen Vernunft* (B), in *Kants Werke. Akademie-Ausgabe*, vol. 3 (quoted from the English translation: Kant 1998).

Kant, Immanuel (1998), *Critique of Pure Reason*, ed. by P. Guyer and A. W. Wood, Cambridge: Cambridge University Press.

Kim, Alan (2015), "Johann Friedrich Herbart", in: *The Stanford Encyclopedia of Philosophy* (Winter 2015 Edition), ed. by E. N. Zalta, https://plato.stanford.edu/archives/win2015/entries/johann-herbart/, visited on January 15, 2020.

Kries, Johannes von (1882), "Über die Messung intensiver Grössen und über das sogenannte psychophysische Gesetz", in: *Vierteljahrsschrift für wissenschaftliche Philosophie* 6, p. 257–294.

Landerer, Christoph/Huemer, Wolfgang (2018), "Johann Friedrich Herbart on Mind", in: *Philosophy of Mind in the Nineteenth Century*, ed. by S. Lapointe, London: Routledge, p. 60–76.

Leary, David (1980), "The Historical Foundations of Herbart's Mathematization of Psychology", in: *Journal of the History of the Behavioral Sciences* 16, p. 150–163.

Maier, Anneliese (1968), *Kants Qualitätskategorien*, *Kantstudien* Ergänzungshefte 65 (1930), repr. in: *Zwei Untersuchungen zur nachscholastischen Philosophie*, Rome: Edizioni di storia e
letteratura.

Mally, Ernst (1904), "Untersuchungen zur Gegenstandstheorie des Messens", in: *Untersuchungen zur Psychologie und Gegenstandstheorie. Zum zehnjährigen Bestande des Psychologischen Laboratoriums der Universität Graz*, ed. by A. Meinong, Leipzig: Barth, p. 121–262.

Marek, Johann (2019), "Alexius Meinong", in: *The Stanford Encyclopedia of Philosophy* (Spring 2019 Edition), ed. by E. N. Zalta, https://plato.stanford.edu/archives/spr2019/entries/meinong/, visited on January 15, 2021.

Martinelli, Riccardo (1996), "Il problema delle grandezze intensive nella filosofia dopo Kant", in: *Rivista di filosofia* 87, p. 445–471.

Martinelli, Riccardo (1999), *Misurare l'anima. Filosofia e psicofisica da Kant a Carnap*, Macerata: Quodlibet.

Martinelli, Riccardo (2002), "Kant, Mendelssohn e l'immortalità dell'anima", in: *Studi Kantiani* 15, p. 93–126.

Martinelli, Riccardo (2020), "Meinongian Psychology", in: *Psychological Themes in the School of Alexius Meinong*, ed. by A. Dewalque and V. Raspa, Berlin: De Gruyter, p. 11–32.

Meinong, Alexius (1896), “Über die Bedeutung des Weber’schen Gesetzes. Beiträge zur Psychologie des Vergleichens und Messens”, in: *Alexius Meinong Gesamtausgabe*, ed. by R. Haller et al., vol. 2, Graz: Akademische Druck- und Verlagsanstalt 1971, p. 215–376.

Meinong, Alexius (1899), “Über Gegenstände höherer Ordnung und deren Verhältnis zur inneren Wahrnehmung”, in: *Alexius Meinong Gesamtausgabe*, ed. by R. Haller et al., vol. 2, Graz: Akademische Druck- und Verlagsanstalt 1971, p. 377–471 (quoted from the English translation: Meinong 1978, p. 73–135).

Meinong, Alexius (1978), *On Objects of Higher Order and Husserl’s Phenomenology*, ed. by Marie-Luise Schubert Kalsi, The Hague: Nijhoff.

Mendelssohn, Moses (1767), *Phädon, oder Über die Unsterblichkeit der Seele, in drey Gesprächen*, Berlin/Stettin: Nicolai; also in: *Gesammelte Schriften Jubiläumausgabe*, repr. Hamburg: Meiner 1979.

Potrc, Matja/Vospernik, Miklavz (1996), “Meinong on Psychophysical Measurement”, in: *Axiomathes* 1/2, p. 187–202.

Russell, Bertrand (1899), “Review of Meinong, Über die Bedeutung des Weber’schen Gesetzes”, in: *Mind* 8 (n.s.), p. 251–256.

Russell, Bertrand (1903), *The Principles of Mathematics*, Cambridge, Cambridge University Press.

Russell, Bertrand (1905), “On Denoting”, in: *Mind* 14 (n.s.), p. 479–493.

Seron, Denis (2011), “The Fechner-Brentano Controversy on the Measurement of Sensation”, in: *Revue Roumaine de Philosophie* 55, p. 87–102.

Smith, Barry (1996), *Austrian Philosophy. The Legacy of Franz Brentano*, Chicago: Open Court.

Sturm, Thomas (2009), *Kant und die Wissenschaften vom Menschen*, Paderborn: Mentis.

Tegtmeier, Erwin (1996), “Meinong on Measurement”, in: *Grazer philosophische Studien* 52, 1996/97, p. 161–171.

Trincker, Dietrich (1969), “Hering, Ewald”, in: *Neue Deutsche Biographie* 8, p. 617–619, https://www.deutsche-biographie.de/pnd119220563.html#ndbcontent, visited on January 15, 2021.

Turner, R. Steven (1994), *In the Eye’s Mind: Vision and the Helmholtz-Hering Controversy*, Princeton: Princeton University Press.

Woodward, William R. (2018), “Fechner (1801–1887). For and In Psychology: Part 1”, in: *Archives of Psychology* 2, p. 1–21.

Katherine Arens

Pedagogy as Epistemology: Building the Subject of Knowledge

Abstract: "Pedagogy as Epistemology" addresses the work of Johann Friedrich Herbart in pedagogy with two goals in mind. The first is speculative, tracking why and how Johann Friedrich Herbart's work could become the official pedagogy of the Austro-Hungarian Empire, while that Austro-Hungarian reading of Herbart was totally effaced in his reception in the United States. The second goal is to recoup Herbartianism in Austria-Hungary as a tradition lying far from today's images of Herbartianism in the US and beyond. This essay begins with Gustav Jahoda's reclamation of Herbart for social psychology, especially in his focus on the collective mental work of the group. Thereafter, it tracks differentiations between Herbart and the Herbartians who codified his work in ways leading to the movement's eclipse, and finally to a different reading of Herbart's pedagogy, as implemented in the 1849–65 school reforms in Austria-Hungary. My results account for the subject of knowledge as part of a communication community and as an embodied cognitive agent, arguing Herbart as providing a vision of education that clarifies what is at stake in understanding three generations of "Austrian" philosophy and science as based on a critical intelligence providing ongoing critiques of knowledge production in all its forms.

Introduction

This essay speculates about why and how Johann Friedrich Herbart's work could become the official pedagogy of the Austro-Hungarian Empire, and how that Austro-Hungarian reading of Herbart was essentially effaced in his reception in the United States.

That fate remains the ground for a historical "truth" centered particularly in the Anglophone world: a fundamental misunderstanding of Herbart's pedagogy that insists on its alignment with morality-based teaching. Yet that "truth" is a factor only in what must be called the "Herbart Industry" in the US, based on a particular German-American inheritance. In Austria-Hungary, the Herbartianism

Katherine Arens, The University of Texas at Austin
arens@austin.utexas.edu

https://doi.org/10.1515/9783110747324-005

of official state pedagogy found in Herbart's work what best might be termed a critical epistemology: a strategy of teaching that revealed knowledge neither as ideological constellations nor as absolute truth, but rather as a body of situated truths, preserved in inherited concepts and often needing revision. This Herbartianism was well aware of the force of what we would call today master signifiers in the collective mental work of the group, both for good and ill, and it insisted on understanding knowledge as needing to be understood in its limits as well as its virtually unquestionable core elements.

To make this case, I will start with a particular reception of Herbart's work that lies outside of education proper – an essay by Gustav Jahoda, an Austrian-born and partially British-educated specialist in cross-cultural psychology, one of the founders of the European Association of Experimental Social Psychology. His work outlines how Herbart clearly tied education to the purposes of the state, particularly in fostering a particular critical epistemology as his pedagogy, but not as an absolutist demand.

After that, I will compare the two Herbartian initiatives above, one in the German-American sphere, and the other in Austria-Hungary, in light of Herbart's own texts and the accounts given of it – in academic texts codifying the Herbart Industry for education, as well as in the practical texts brought into life by the great school reform of 1849–1856 in Austria-Hungary. The result will be an account of the subject of knowledge as part of a communication community and as an embodied cognitive agent – a vision of education that clarifies what is at stake in understanding three generations of "Austrian" philosophy and science as based on a critical intelligence providing on-going critiques of knowledge production in all its forms.

1 Jahoda's Herbart: From Psychology to the State

In a 2006 article entitled "Johann Friedrich Herbart: *Urvater* of Social Psychology", Gustav Jahoda (1920–2016)[1] makes the case for Johann Friedrich Herbart's centrality for the history of psychology as he moved beyond individual psychology and toward social psychology. Trying to rescue him from Anglo-American neglect in the history of psychology (Jahoda 2006, p. 19), Jahoda traces Herbart's

1 Forced into emigration after being expelled from the *Gymnasium* in 1938 because of Jewish ancestry, Jahoda studied sociology and psychology and took up a post in Ghana in 1952, then was called back to Strathclyde in 1963, where he finished his career. He published works on cross–cultural psychology, socio–cognitive development and history of the social sciences – over 200 articles in all. His honors were numerous (Deręgowski 2017).

(familiar) long-lasting influence in nineteenth-century psychology and international pedagogy. More important for the question of what that influence consisted of is Jahoda's proof that Herbart not only established psychology as a natural science, but also extended "his individual psychology to social and political issues" over the concepts of group, (local) community, and society (*Gruppe, Gemeinde, Gesellschaft* [Jahoda 2006, p. 20]).

Jahoda tracks Herbart's use of an underlying metaphor of communication, wherein each individual is bound into a group, and hence into a group psychology that grounds them as what we today might call an epistemic community, as subjects who espouse a common purpose, as citizens, with motivations extending beyond boundary of the state and the power involved in it (Jahoda 2006, p. 24f.).[2]

Here, Jahoda is consciously countering conventional narratives about Herbart, in insisting that he is interested in the group, not just speaking as a moralist. Those conventional narratives usually also source that moralizing to Bolzano's deontology, rather than Kant's categorical imperative: the former, more norm- or virtue-centered, and the latter, more agent- or practical-virtue-centered (dealing with individuals' *choices* of acting according to the good). Yet "deontology" is a term that has more flexible roots: deriving from the *deon* (the Greek word for duty) and *logos* (which meant simply reasoned argument, in Aristotle's rhetoric); it is not referenced to any absolute Platonic norm, but rather into a communication community. Only in contemporary moral philosophy does deontology become normative in comparison to Aristotelian virtue-ethics (see Alexander/Moore 2016). Jahoda himself distinguishes Kant and Herbart in this way:

> The relation between Kant's and Herbart's theoretical positions was a complex one. There is a sense in which Herbart depended on Kant, taking over many of Kant's ideas even though he often modified them. However, Herbart distanced himself from Kant's metaphysics in regarding the mind as a dynamic system interacting with the environment. For Kant causation was a mental category, while Herbart treated it in deterministic fashion.[3]

Jahoda's "determinism", however, is not nature, but cultural inheritance.

I would remind my readers that, in the nineteenth century, an alternative existed to make the bridge between metaphysics and historicized, contextualized knowledge communities: Johann Gottfried Herder's ideas of groups evolving

---

2 In more modern terms, drawing on Althusser's idea of "Ideological state apparatuses" that reproduce the state or group's dominant ideologies, one might here point out that ISAs create citizens, whereas common experience generates communities that can reach beyond borders.
3 Jahoda 2006, p. 36.

under pressure of experience, which is itself often considered nationalist-deterministic, but which shows that "determinism" is quite different than "permanence" (especially of ethnic determinism) or stagnation. I have argued elsewhere that this more static determinism is not generally found in discussions from Austria-Hungary (Arens 2016), which read Herder as situating (group) psychology within historically situated groups sharing experience and acculturation. In this alternate reading of his work, Herder cannot be considered an ethnic nationalist in the modern sense. Rather, Herder found determinism only in the group experiences that are passed on to a next generation – in the *Vorstellungsmasse* –, not in any ethnic-genetic necessity, which is confirmed in another way in Herder's work on aesthetics, documenting his awareness of plasticity in mind and ideas.[4] It is the topic of another discussion to argue that Kant, too, was much less the cognitivist than what Jahoda claims, but that a more anthropological Kant is straightforwardly recoverable, as well, especially from his shorter essays that philosophy often ignores in favor of the *Critiques*.

Jahoda notes that purported differences between Kant and Herbart loomed large in Herbart's era, which emerged in no small part because of their different approaches to psychology. Kant took physics (and the physics of forces) as the basis for his model of mind (Jahoda 2006, p. 23), where Herbart included biology. More significant for the transition from psychology to education that is the topic here, however, is that *both* used metaphors (either biology or physics) that implicated mechanical force to representations – explaining how representations, in a mass configuration integrated in the mind (*Vorstellungsmasse*), exerted force (*Kraft*) or pressure so that shifts in concepts also require energy, as concepts actively resist each other in a quasi-mechanical fashion. In one sense, this is basic physics, applied to the unseens of represented concepts: bodies at rest tend to stay at rest. Yet Herbart made an explicit transition in establishing connections between bodies and mental representations,[5] insisting that variations in concepts came into being because of local environments (Jahoda 2006, p. 25), and that representations come in with the mother tongue, as philologists since Herder and Humboldt have acknowledged.

Herbart had a second analogy drawn from physics that also recurred as he was building up the picture of mind in his various works on psychology, where he outlined the relationship between psychology and the social sciences, es-

4 This is the case of Herder's discussion of dialects in the second part of the *Abhandlung über den Ursprung der Sprache*.

5 In his "Pestalozzis Idee eines ABC der Anschauung" (1802), Herbart already insists that observation of the visible and physical world is the first step in learning.

pecially in the form of politics science (*Staatswissenschaft;* Jahoda 2006, p. 26). He describes both minds and disciplines as *systems* and insists that every system, be it a mind or discipline or even a state form, has a statics and a mechanics (a mechanism) (Jahoda 2006, p. 31). These systems also function in terms of energetics, with strong concepts creating social bonds and weaker ones signaling less dominant concepts and positions within a group. Additionally, Jahoda does us the service of making the bridge from psychology to his version of social psychology over the idea of *Volksgeist* ("the essence of the people", a concept originating in the post-Humboldt linguistic thought of Lazarus and Steinthal [Jahoda 2006, p. 34ff.]).

From today's perspective, Jahoda's Herbart is an early entry in the kind of idealism that connected groups and state forms, one that is also found in the work of Johann Gottfried Herder, where it derives from the late eighteenth- and early nineteenth-century work on language and mind. This well-documented "interdisciplinary" tradition actually rests on a more nuanced reading of Kant, on a model of mind creating knowledge out of world experience, which implicates it with historicism as well as communication communities, based on individuals' shared experience of history, and also on historical transmissions within groups, and, of course, in the group's language. In this, Jahoda opens out the historical epistemology surrounding Herbart's work, tying it to a series of texts explicitly addressing language and culture, from Johann Gottfried Herder's *Essay on the Origin of Language* (*Abhandlung über den Ursprung der Sprache*, 1769) through the work of Wilhelm von Humboldt, and then Heymann Steinthal and Moritz Lazarus. Following this thread into what has come to be identified as social sciences, we see an evolving set of projects, now identified as separate disciplines, that start out by tying mind to communities situated geographically (as different anthropological projects) and historically, and constantly under pressure by the groups' experiences.

Jahoda realizes that focusing on a mass of concepts as key to national identity (not ethnic-genetic but based on experience and tradition) also places Herbart's work at the originating intersection of the social sciences as we know them (and that had not yet broken apart from each other). More particularly, it builds the state as an image of a community developing from below rather than fulfilling a providential promise made within creation and sent from above – an image of the state and its power foreshadowing the work of Michel Foucault in highlighting cultural inheritance as a power-knowledge network.

What Jahoda does not stress, however, is that, from a more practical standpoint, this tradition was also involved from the first with pedagogy as an ideological state apparatus (again, Althusser's terminology) because it describes individual minds as under the influence of culture (language, concepts, traditions) as well as nature (experience through the body). As such, the "little brother" discipline evolving from Kant's Copernican revolution, alongside psychology and linguistics (philology), and anthropology, was *pedagogy*: how to *form* the individual within the group, and to integrate that individual into a collective consciousness that recognizes and works toward some greater good. Here, Jahoda's "social psychology" must necessarily also be seen as implicating state interest, as the emerging modern nation-states of Western Europe (and later large parts of the world), took up the challenge of education in the service of the true and the good – and the politically necessary.

And here, we reach a partial cause of Herbart's effacement from "modern" histories of Western states and the disciplines in them. As education became associated with the state and how it controlled knowledge, rather than being an issue of how individuals generated knowledge, it did so in an era which posited a political break between the Kantian-Herder lines of idealism I have been positing here, read as fiercely committed to knowledge through experience within local communities. This tradition, and Herbart, fell out with the Hegelian-Romantic line of idealism that was increasingly associated with nationalism and ideas of individual genius. This later idealist tradition not only fed ideals of national character and destiny as somehow part of creation rather than of historical memory, but also led to a hardening of the terminology that characterizes German thought from Romanticism onward: a break between *Erziehung* and *Bildung*, with the former coming to mean school education (literally, in its linguistic roots, up-bringing) and the latter the formation of character (the Romantic fallacy of genius and ability through birthright, which needs individuals' direct engagement with the world, rather than school-learning, for the potential innate in individuals to emerge as genius or in the form of world-historical individuals).

2 Herbart between Pestalozzi and the Herbartians

At this point, any discussion of Herbart as contributing to education and to ideas of community runs up against a profound problem in considering philosophies of education and their realizations in actual school curricula: namely, *Herbart's* work is by no means equivalent to that of the *Herbartians*, who emerged at full strength and with recognizable organized practice in colleges and universities only about 25 years after Herbart's 1841 death. Where Herbart considered education as relating to both the mind and community as changing through history, the Herbartians all too often became normative and overtly moralistic.

In German, there is a general term, *Reformpädagogik*, which refers specifically to school reform, teaching and learning, starting with the child who will become the adult and citizen. This tradition is reconstructed as reaching back to Comenius, Rousseau, and particularly Johann Heinrich Pestalozzi (1746–1827);[6] its history is told in the form of competing models about what and how children should learn as they mature toward adulthood, and, as such, the discipline as presently understood is associated with one kind of reform pedagogy and is often accused of being abstract and philosophical-abstract.

In general terms, a "reform pedagogy" has as its goal service to society: the need to *improve* and *expand* the percentage of the population that gets educated, and to clarify what each group to be educated needs and how those needs can be transformed into practice. In one sense, this tradition may even be traced back to the state-threatening pedagogy of the "Little Schools" of Port-Royal-des-Champs, a Jansenist abbey whose schools were first closed by papal bull in 1660–61 and whose pilot institution, an abbey of nuns, was then forbidden by another papal bull in 1708, cleared out by Louis XIV in 1709, and its buildings razed. Still, *Port Royal* and its textbooks have remained a central memory in explaining the emergence of the *Lumières*, especially those "Enlighteners" working on the *Encyclopédie* and from non- or lesser-noble families. Even today, many progressive primary schools in the Federal Republic of Germany are still *Pestalozzi-Schulen*.

If the Jansenist pedagogies were radical, challenging the authority of the state by purportedly questioning the absolute authority of the upper classes in their choice of educating individuals of many classes and origins, the other end of the continuum were the latter nineteenth-century *Herbartians*, a second group of self-proclaimed *Reformpädagogiker* who became the state employees opening schools purportedly based on Herbart's work (or marketing curricular innovations for sale in the form of textbooks and learning aids [today's educational toys]); other such reformers extended Herbart's institutional legacy by expanding the reach of teachers' seminars, creating a new discipline, but not always a climate for reform.

In the United States, perhaps the largest consumer of reform pedagogy outside of Central Europe (and with a large German-American population who *studied* reform pedagogy there), another term arose: "progressive education", which set itself apart from the older European traditions like the English public schools and the German *Gymnasium* by deemphasizing the classical humanist curricula aimed at replicating class positions and preparing students for the university.

6 Swiss educational reformer known for a pedagogy associated with experiential learning and observation and summarized by a motto: "Learning by head, hand and heart."

Progressive education in the US today is associated with Montessori schools at the primary level and aligns closely at that level with the philosophies of education that Herbart knew, especially Pestalozzi's. This education is associated with fostering and strengthening individual learners through education suiting their own cognitive styles.

One familiar example shows, however, how such pedagogies can be monetized: still today, one can buy kits of the "Froebel Gifts" that were to foster children's preschool and primary school learning through hands-on learning toys and tasks.[7] These have an analogue in the Montessori sensorial materials used in early years of these schools. The toys are different in detail of implementation but share the purpose of encouraging, through hands-on experiments with physical objects, not only play, but also early problem-solving, manual and cognitive dexterity, and critical thinking.[8]

By the time progressive education becomes a watchword in the US, early applied to schools like the University of Chicago Laboratory Schools (founded by John Dewey in 1896) or the Francis W. Parker School (founded in 1901, also in Chicago), the term also implies a unified curriculum articulated across levels, including group work by students (often in a thematic focus that allows individuals to contribute to a whole according to their own interests and abilities), formative assessment, and an integration between school and the needs of (a democratic) society – a strong overtone of egalitarianism instead of elitism.

However, these virtues for an ideal "reform" pedagogy in the United States were claimed by a nativist move: by derogatory reference to US Herbartianism (associated with Charles De Garmo, Charles Alexander McMurry, and Frank Morton McMurry, who all studied at either Halle, Jena, or both), which had evolved as derived from the post-Herbart German Herbartians, especially Tuiskon Ziller (1817–1882, at Leipzig), Karl Volkmar Stoy (1815–1885, at Jena), and Wilhelm Rein (1847–1882, at Jena). They were later branded by the progressive education movement in the US as dogmatic moralists who constructed rigid, unyielding pedagogical schemes as their "curricula" that they taught at US institutions. In this judgment (and in the advertising of "US educational innovation" like Parker and Dewey), German reform pedagogies became branded as prescriptivist. Yet just as the German Herbartianism held pride of place in the Universities, their first US students created the curricula of the first US teacher's colleges, then called "normal schools". Here, a structural difference arose: US normal schools taught

7 Friedrich Wilhelm August Fröbel (1782–1852) was a student of Pestalozzi who was a pioneer in the Kindergarten movement.

8 Pictures of both standard sets are available in the Wikipedia articles under their names.

primary school teachers, where European pedagogical seminars taught teachers for both primary and secondary education, and sometimes into post-secondary levels (hence the French term *Écoles normales supérieures*, which often no longer teach pedagogy at all) – the terminology is in flux, but the major differences ended up again being distinctively class-based, with teacher-training progressively concentrated in public schools in the US. "Normal" schools were dedicated to "standardized" education and predictable educational products, including the inculcation of "national values". It was therefore straightforward for "reformers" like Dewey to appeal to Anglo-American ideals of the elite as somehow lying outside of these institutions, if they persisted.

Looking behind this marketing, however, reveals a quite different reality that again separates *Herbart* from later *Herbartians*. Yet one should not forget that Herbartism's narrative relies not just on the shared inheritance in the teachers' seminars in Germany where many of these US pedagogues studied before they produced their own textbooks, but also their experimental schools and the social experiments of their advertising. The original German notion of a *Reformpädagogik* looked very much like the US laboratory schools that continue to this day, not like the moralistic and closed-ended curricula attributed to the US Herbartians (see Cruikshank 1998; Somr/Hrušková 2014, especially p. 417ff.). Although there were many variants, the fundamental model for US-Herbart curricula actually resembled nothing so much as an early form of what we would become familiar with in the 1950s as *Bloom's Taxonomy*, moving from comprehension to active association/construction of knowledge, and from there to applying that knowledge.

The German terminology ranges from "Klarheit, Assoziation, System und Methode" (clarity, association, systematizing, and method) to Wilhelm Rein's (1847–1929) "Vorbereitung, Präsentation, Umgang, Verallgemeinerung und Anwendung" (preparation, presentation, practice, generalization, and application) as the description of the sequence of activities that foster learning (see Somr/Hrušková 2014). In practice, this meant that the learner was first oriented to the materials to be studied in presentations by the instructor, then was brought to make connections between these materials and facts established out of their own material experience, working toward constituting the two sources into a system or system of meaning. At the highest level of achievement, learners were to apply that systematic comprehension to new situations or materials.

Unfortunately, that *strategy* for moving individuals toward independent inquiry soon became part of an evolving set of increasingly normative theories that *prescribed* how students were supposed to be approached. For example, early Herbartians only knew four steps: *Klarheit, Assoziation, System und Methode*. In the late 1850s, Tuiskon Ziller (1817–1882) added "Analyse" to that list (one specific kind of preparation, sometimes referred to as "correlation", to highlight evoking student interest in the unfamiliar by linking it to the familiar; he was interested in education as *recapitulation* of development). It was Wilhelm Rein (1847–1929) who changed the terminology into something closer to the modern form: *Präsentation, Umgang, Verallgemeinerung und Anwendung* (see Müller 2000 for an overview). Herbart, however, wanted the students to develop their own personalities as flexible, multi-faceted learners and moral social agents,[9] not to be forced into single mode (which later pedagogues like Maria Montessori would stress as undesirable uniformity). In using the idea of character, in addition, Herbart was stressing a property of will (not innate self) – the *willingness* to engage with new knowledge and experience, including the input from the community and inherited culture.

As this structure evolved, Herbart's framework became codified in ways he had not intended, as well. Attributed to Herbart himself in the various pedagogical treatises written after his death is a call to frame education in three ways, according to *Formalstufen* (stages of increasing formal difficulty), *Kulturstufen* (the stage-wise evolution of cultures through history), and the *Konzentrationsidee* (student-centered focus on enhancing attention and motivation), all resting on the idea that the student can be transformed through education. It was only the "formal stages" that became codified as: preparation, presentation, association, generalization, and application, which prescribed an order to classroom practice. These formal stages became a grid for lesson plans on how a teacher was to introduce materials: first, in relation to existing knowledge, and then concretely, then compared with prior knowledge in order to facilitate the development of more general (and abstract) principles, and finally applied independently. This provided a grid based on induction (from the concrete to the abstract) that could be used by teachers as a guide to lesson planning, and by administrators for evaluating teacher classroom practice. Moreover, this fundamental model for US-Herbart curricula actually resembled nothing so much as an early form of what we would become familiar with in the mid twentieth century as Bloom's *Taxonomy of Educational Objectives: Handbook I: Cognitive* (1956), moving from comprehension to active association and from there to applying the new knowledge.

The Herbartians' *concentration* turned into a focus on motivation, where it had been cognitive engagement and interest in Herbart. Their "stages of culture" all too often became identified with a normative reading of historicism in

9 This reading is generally based on *Die ästhetische Darstellung der Welt als das Hauptgeschäft der Erziehung* (1804).

nationalist perspectives, where the original stressed history not as a norm, but in order to structure a sequence of presentations that would conform to pupil's age-appropriate abilities of conceptualization and systematization.

It remains a critical difference that Herbart himself adhered to the spirit of reform or progressive education, focusing not on normative learning goals but on shaping the individual learners' will and experience. He did eventually open one of Germany's first teacher's seminars at the University of Königsberg between 1809 and 1833. Yet he was involved both in traditions of theory and praxis alike that largely refused to assume uniform educational outcomes, instead of encouraging practices of mind that lead individuals to produce knowledge.[10] Remember that Herbart came to pedagogy not only through his family (his grandfather was a pedagogue), but also when he was a tutor in Switzerland (between 1797 and 1800). There, he became familiar with the work of Johann Heinrich Pestalozzi (1746–1837), the most famous and established philosopher of education in the era (after Rousseau), an educator and paradigmatic school reformer of the late eighteenth and nineteenth centuries. Pestalozzi was best known then from his novel of pedagogy, *Leonard und Gertrude* (first volume in 1781, reaching four volumes by 1785), which described a utopia of family- and community-based education, continued in *How Gertrude Teaches Her Children* (1801). Pestalozzi also tried to realize his principles in a primary school in Burgdorf that was a success, but less so when transferred to other locales.

Herbart and Pestalozzi enjoyed from the start of their personal acquaintance a complementary relationship, not a dependency. The story is that Herbart recommended a French tract on education to Pestalozzi which convinced him that psychology was also a necessary part of education, in prescribing what we would call today a learning sequence – usually called an *Associationspsychologie* and tied into something like Jean Piaget's work on child development today. Yet Herbart had also been set on this course by his own studies. He had started university studies in Jena under Fichte, pursuing a law degree. His interlude as a tutor was the by-product of quitting law studies and leaving Switzerland in 1800; he reentered university at Göttingen in 1802, where he finished his degree and *Habilitation* by 1805. Herbart's career was on a rising arc, as he was called to a chair at Heidelberg at 1806, and then, in 1809, to Kant's chair at Königsberg,

---

10 Herbart was in 1809 called to Königsberg to Kant's chair as a *Professor für Philosophie und Pädagogik*, in which position he wrote some of his most important books: the *Lehrbuch zur Psychologie* (with an 1816 first edition with several others following) and the 1824/1825 *Psychologie als Wissenschaft, neu gegründet auf Erfahrung, Metaphysik und Mathematik* in two parts, as well as several other books on philosophy.

where he was active in Prussian school reform (and was noticed by Wilhelm von Humboldt, himself the reformer of the Prussian university system). Herbart held the Königsberg chair until 1833, when, failing to secure Hegel's chair in Berlin, he returned to Göttingen, where he stayed until his death.[11]

Yet as school reform became the West's project for the nineteenth century under the rubric of reform pedagogy or progressive education, the student-centered core of the educational model that Herbart shared with Pestalozzi all too often became conflated with the particular moralities of the nations and cities that adopted it – for every true reform pedagogy, there was a more dogmatic one innovation driven by national projects. Herbart's insistence on character was often used to justify education to be pushed into the service of particular moralities, as opposed to an education that strengthened the learner's will in understanding and applying *values* in the sense of Kant's second *Critique*: fostering the ability to judge in moral-valuative dimensions.[12] Thus Herbart's student-centered ideals, tempered with an association with community values), were too often implemented in a way easily derided as teacher-centered, in which the teacher, as an employee of the state, sets the rules and the hierarchy in inherited terms. US Herbartianism became associated with a limited moral education, focused on values like Inner Freedom, Perfection, Benevolence, Justice, and Equity or Recompense. In reflection of a US educational establishment that stressed schooling for "daily living" (see Blyth 1981).

To correct this skew, it is again helpful to turn to the original texts to see how *community*, not just individual talent or national interests (defined abstractly), is implied by Herbart's *Reformpädagogik* – and ultimately to see how that definition inspired a generation of pedagogues in Austria-Hungary in a school reform that was designed to empower minds as part of knowledge communities.

If Jahoda showed us the historical-communitarian dimensions of Herbart's contribution, the *Stanford Encyclopedia of Philosophy* can serve as a thumbnail of a different understanding of Herbart, returning his work to the context of a philosophical psychology of cognition. As it summarizes:

---

11 As dean (*Dekan*) of the Philosophical Faculty, he disapproved of the famous 1837 "Göttinger Seven" who protested in writing directly against the king's suspension of an 1833 liberal constitution – Friedrich Dahlmann, Jacob Grimm und Georg Gottfried Gervinus were the three of the seven who were banished from Hannover, only to be welcomed into Prussia by 1840.

12 Here, we see again a connection to Dewey, who insisted on education to citizenship – which supposedly the Herbartians were not doing (in their lockstep lessons that were more generally western and humanist in inspiration). Herbart's "ethics" conforms more closely to a reading of Kant's categorical imperative that situates norms within communities rather than as *a prioris*.

> Instruction aiming at many-sidedness has two main phases, "engrossment" (*Vertiefung*) and "reflection" (*Besinnung*), each of which in turn has a "resting" and a "progressive" moment: resting engrossment is "clarity"; progressive engrossment is "association"; resting reflection is "system"; progressive reflection is "method".[13]

This phrasing of the English translation still largely ignores the German terminology, but it does capture a nuance by accident or design that points to communities of understanding in divergent ways. The term "engrossment" today suggests only part of the word's historical definition: the fact of being enthralled, mesmerized, or held rapt by a thought or object. But in legal terms (which would have been familiar to the earlier translators), "engrossment" is also a legal term, referring to the process of preparing a document for final printing or drafting on to fine paper for signing – the fine draft of what will become a binding contractual obligation, as in terms of the state. This cross-reference would have well-suited US Herbartians and their education toward community norms.

In German, however, *Vertiefung* implies not just enthrallment or a contract, but rather absorbing one's self into, delving, or deepening one's encounter with a subject matter, in the sense of extending knowledge and going into depth on a subject (*vertiefen* as a verb, with *sich vertiefen* meaning to immerse one's self into something) – a metaphor of expansion, rather than a contract. Going into depth had "resting" and "progressive" phases, corresponding roughly to the comprehension and active use of new concepts. The "clarity" sought in such an exercise includes recognizing novelty or discovery, but not creation: the learners are to integrate, not innovate – simply to understand how the new connects to the old, their *Vorwissen*. The active side of *Vertiefung* is *association*, a term familiar from eighteenth-century psychology. Active *Vertiefung* requires individuals to do the work of immersing themselves into the data and enmeshments of the concept and impressions and consolidating them into the mass of ideas already present.

Besinnung, reflection, is a term most famous from Johann Gottfried von Herder, who states in his *Essay on the Origin of Language* that what separates the human mind from animals is the aptitude for *Besonnenheit*, the ability of an individual mind to reflect on impressions. Here again, the English is a bit misleading: through reflection, one first inserts new knowledge into the field of old knowledge, and then practices that knowledge (*einüben*, a verb suggesting automaticity).[14]

---

13 URL = https://plato.stanford.edu/entries/johann-herbart/section 5.1.

14 The difference between *üben* and *einüben*, which both mean "to practice", is significant here: the pair moves toward the distinction between "practice" and "rehearse" in music, with the latter referring to a more active command of the materials.

These distinctions may be traced back to the nineteenth-century practice in psychology of distinguishing "interest" and "association" that derive from Kant's notion of *Anschauung*, the immersion of the self in the physical. But Herder's ideas about language again add an interesting dimension to this adaptation: that the first words that originated are each a *Merkwort*, a designation that calls attention to a group of perceptions, focusing attention. That *Merkwort* turns ideas into the concretes – to spoken words – where it can be used to transform the communication community once it is *grasped* (*begriffen*, a verb associated with *concept/Begriff*). *Interest*, then, is not just cognitive, it is aesthetic, passed on into the community, and organized through the work of more active powers of mind – Kant would have called it *productive Einbildungskraft*, the imagination that produces things. Experience is grasped by being taken up and given concrete form, then integrated into a person's knowledge, and then organized.

This sequence was by 1900 known as "Herbart's rule", a set of strategies that need to be tended to in any pedagogy. A period account summarizes that this pedagogy hoped

> to give equal right to engrossment and reflection in every area of inquiry, no matter how small; i. e., to tend in order to: *clarity* of the particular; *association* of the many; *coordination* of what has been associated, as well as practice in progressing *through* this organization. In all this, the elements of instruction – "facts, information, ideas, knowledge" – are not negligible, but rather vitally important, for [...] they become built up into "apperception masses", which, in the process of taking in *more* facts, information, ideas, knowledge, give rise to "apperceptive interest", this latter being itself of first importance in the character-forming process.[15]

A concept, then, needs Herder's *Merkwort* to become concrete in Herbart's sense. And as concrete and integrated into the *Vorstellungsmasse*, the representations in individual minds or the *Begriffe*, the concepts used in concrete contexts, have two sides, one object and one subjective – and both individual character and the character of national "mind" come out of it, stabilized by individual memory or cultural tradition that directs attention to it. The result is a culturally based habit (very much like Bourdieu's *habitus*) that relates the individual to the group. Morality is part of that in the sense of *judging* to standards, not of innate values. The goal is to shape the learner's will to standards of clearness and distinctness – to become an adult in society who will use the force inherent in the *Vorstellungsmasse* to shape and reshape knowledge in light of the group. The teacher thus

15 Hayward 1907, p. 70, cited in Kim 2015.

"forms character" based on individuals' perceptions in iteration with group norms, not just those conforming to those norms.

The teaching in US-style Herbartianism prescribes a goal set by institutions; Herbart, like Herder, prescribes only a logical course of activity building knowledge – setting up the conditions under which credible judgment emerges. This Herbartianism is *teacher-centered* (focusing on a teacher forcing students through these stages of learning in "proper order". Herbart's model of education is *learner-centered,* focusing on active engagement with the learning process – it resists a highly structured curriculum and instead favors something like we call today "readiness", much as Montessori or Waldorf (Steiner)[16] curricula do: they stress human cognitive development rather than the specific contents to be mastered, and the engagement of both mind and body.

These distinctions between Herbart's model of an education based on cognitive psychology and one defined by Herbartian history as institutionalized in the German-inspired pedagogies in the normal schools are, however, not the only options. In fact, Herbart inspired one of the most important educational reforms of the nineteenth century, one that influenced two generations of intellectuals (at least) and implemented a kind of critical thinking on an unprecedented scale.

3 Pedagogy as a Critique of Knowledge Production: Austrian Herbartianism

The systematic application of an educational program somewhat closer to Herbart's philosophy was found in the Austro-Hungarian Empire, on a course defined by philosophy professor Franz Serafin Exner (1802–1853), who brought Herbart's philosophy to the Charles University at Prague between 1832–1849. That pedagogy became the official Austro-Hungarian approach to education when one of Exner's Prague students, Leo Graf von Thun-Hohenstein came into the Ministry for Education (*Unterrichts-Ministerium*) in 1848 as a Minister, where he was put in charge of a commission for reforming the curriculum, especially at the university level. Exner was part of that commission between 1849 and 1851 (see Aichner and Mazohl 2017), having written a set of guidelines for a future reform of the university system in 1844, which had led to nothing in the political unrest of the

16 For an example of a Steiner Curriculum, see http://www.steinereducation.edu.au/curriculum/steiner–curriculum/; for the Montessori and Waldorf movements, see their websites with their curricular frameworks URL = http://www.montessori.com/montessori–method/ and URL = https://waldorfeducation.org/waldorf_education/curriculum (all visited on April 12, 2021).

Vormärz. The plan for the *Gymnasien* (secondary schools) was provided by Hermann Bonitz (1814–1888), a classical philologist from Berlin.

The writing of what would become the "Provisorisches Gesetz über die Organisation der akademischen Behörden" ("Provisional Law on the Organization of Academic Authorities") was completed quickly: it went into effect in September, 1849, with imperial approval, but with the actual reforms being rolled in gradually up through 1856. The resulting plan included momentous changes, including an expansion of the *Gymnasium* to eight years, a plan for *Realschulen* that would allow transfer to the universities, and for the universities a grant (on paper at least) of a degree of *Lehr- und Lernfreiheit* (self-determination of what is taught and learned).

That the implementation of a philosophical program for education could be fast-tracked in this way probably rested on Austria's internal academic politics. The very appointment of Count Thun-Hohenstein argues for that: he was a Bohemian conservative and strict Catholic who wanted to align the educational system with religious values (9), which could be accommodated by Herbart's insistence on the person as part of a community. Just as importantly, however, he was "anti-Kantian" and "anti-Hegel", in the (Austrian? Bohemian?) tradition of philosophy represented in the work of Bernard Bolzano at Prague and his reform of Wolff's ideas about knowledge.

Today, "Austrian philosophy" is remembered for offering a far-reaching critique of language as vehicle of truth; more properly, that tradition needs rather to be seen as a rejection of the separation between ideas and sensory experience. Rejecting that distinction not only rejects the problems of signification foregrounded in the twentieth century by distinctions between sense and reference. This move also is a tacit rejection of the Revelation paradigm still in place in conservative Protestantisms, because it will, like Kant, not assume that knowledge was revealed "from behind" some kind of wall beyond which we cannot see. Instead, this model reaches back to Leibniz and his *Monadology*, which posits "revealed" human knowledge as part of creation – not *a priori*, but co-extant. That position was strengthened by a July 1798 commission at the University of Vienna that sought to offer instruction about Kantian philosophy there – which was defeated as a dangerous philosophy (that is, as leading to atheism). Bolzano, however, did teach Kant – and was censured for it until a protector advocated for him, which staved off consequences until 1819.

The 1849 documents affirmed Herbartian educational philosophy as nominally commensurate with Thun-Hohenstein's goals, but as doing so in a more modern fashion commensurate with a Bohemian Catholic Enlightenment. One of the final documents outlining how to turn that law into practice was published in 1849, under the authorship of the *Ministerium für Kultus und Unterricht*: the *Entwurf der Organisation der Gymnasien und Realschulen in Österreich* outlines an official school system organized in ways that approximate that of the latter twentieth century. Overall, the document tells us in great detail what students, administrators, and teachers are expected to do in class, including virtually every facet of classroom and school management.

The curriculum had clear guidelines, but it and the school types were explicitly intended to be somewhat flexible (1): for example, by allowing students to shift between *Gymnasium* and the more career oriented *Realschule* at the lower levels of the *Gymnasium* (the *Untergymnasium* [3]). The document explicitly allows for schooling to be conducted in *any* language of a community (as its *Unterrichtssprache* [Austria 1849, p. 19]), not just German, and set two or three languages as the goal for education – living languages, not just classical ones, although Greek and Latin could straightforwardly be justified among the three. On the level of personnel, the *Entwurf* specified that teachers should work with parents, and that home-schooling options were to be allowed (as long as the curricula articulated with the official ones); it also allowed for both public and private secondary schools (Austria 1849, p. 12).

On the more conservative side, the *Entwurf* noted that the schools should not only teach but also "build character" (Austria 1849, p. 7). Thus, classical texts, as well as important works of vernacular literatures other than German, needed to be taught, not just for language practice, but also as

> eine reiche Fülle geist- und charakterbildenden Stoffes in klassischer oder mindestens tadelloser Form darbieten, und auf den Unterricht in sämmtlichen anderen Lehrgegenständen belebend, verknüpfend und theilweise ergänzend wirken.[17]

This would include literature in specified genres from Opitz to Herder (Austria 1849, p. 120), as well the more modern Schiller and Goethe (Austria 1849, p. 135).

Critical for Herbartianism in Austria-Hungary are two additional, simple rules that at first also seem conservative: that approved textbooks must be used in schools at every level, but at the same time, teachers were not simply to repeat the textbooks in class – a decisive move beyond teacher-centered classrooms and the recitation model of teaching (Austria 1849, p. 41). This created an opportunity for classrooms to come to their students.

What is also clear in such dicta is that the *Entwurf* was concerned with articulation across levels, defined in terms of what today might be called cognitive

---

17 Austria 1849, p. 28.

readiness: clear attention was paid to the psychological and social readiness of individual pupils (and here, again, we must be reminded of Gustav Jahoda's insistence on Herbart as a social psychologist). Thus, for example, the sequence of study is defined in distinctly Herbartian language (*Vertiefung*, "empirischen Auffassung des Einzelnen", and students' *Selbsttätigkeit* all appear [Austria 1849, p. 176]). For instance, the kinds of reading texts to be used at different levels were defined in terms of cognitive complexity and readiness. Thus, a reading curriculum starts in the *Realschule* with the epic fable, then proceeds to the epic *Märchen*, and then finally to *Robinson Crusoe* (a series leading from more realistic to more speculative prose texts, and then to literature writ narrowly). The chosen texts often are organized in historical order, but, as in the Vienna School of Art History (associated with Alois Riegl), historicized *orders* of texts were not construed as reflecting *cultural progress* or *masterworks*, as Hegelians might do, but rather more straightforwardly. Instead, texts were associated historically as they reflected a group's experiences – a society with limited technologies and small populations would not have the complexity of modes of representation and signification that were possible in a larger society. This is an acknowledgment of historical distance, rather than historical determinism: "Wie 'Classicität' ist auch 'romantisch' ein systematischer Begriff ohne begrenzten zeitlichen Bezug", notes Jäger (1982, p. 216) – abstract organizers like genre and era are descriptions of particular cognitive organizations within groups, not a state of historical development.

Jäger outlines how this historicized Herbartianism guided by a philosophy of mind was critical for the modernization of education on the basis of psychology rather than on a normative set of national values:

> [Herbarts] Lehre kennt keine Natur- und Geschichtsphilosophie, die die Selbstrealisation Gottes oder die Offenbarung der Idee im Werden der Dinge verfolgt. An die Stelle der geschichtlichen Behandlung tritt in seiner Schule die Psychologie als Grundwissenschaft. Die Bedeutung dieser Tatsache für die österreichische Geistesgeschichte wird m. E. noch nicht voll erkannt; sie eröffnet auch erst das Verständnis der Formalästhetik [...]. Alle äußeren wie inneren Wahrnehmungen sieht Herbart als psychische "Vorstellungen" an, die sich aus dem Zusammenhang ergeben, in dem wir uns als einfache Wesen mit anderen gleicher Art befinden.[18]

The fundamental premise underlying Herbart's work is Kantian: all things known are representations (*Vorstellungen*), set into relationship with each other, and the historical subject of this knowledge is the product of such relationships (Jäger 1982, p. 202). Learning, then, should be structured to proceed inductively, to help

---

18 Jäger 1982, p. 199.

the learners learn to uncover these relationships, as well as to critique and to expand their knowledge of them. Moreover, that learning process also pays attention to feelings and experience, not just logic and concepts: "Gefühl ist das Bewusstsein der Spannung von Vorstellungen" (Jäger 1982, p. 211).

These principles for structuring curricula resonate strongly on Herbart's original synthesis, written as early as his first Göttingen era, such as *Über die ästhetische Darstellung der Welt als das Hauptgeschäft der Erziehung* (on the aesthetic representation of the world as the main concern of education), published in 1804, and *Allgemeine Pädagogik aus dem Zweck der Erziehung abgeleitet* (general pedagogy deduced from the aim of education), published in 1806. Both also have a "moral" dimension that, however, is not moralizing – he uses the term *Lebensordnung*, the ordering of life as lived, as his goal. Here again, we might see this as a kind of Aristotelianism of practical living rather than a belief in enduring values.

The Austro-Hungarian laws for school reform opened the door for at least two generations of textbook writing that are the basis for the true reform pedagogies familiar from the twentieth century – not only to Pestalozzi and Montessori grade schools, but also to the secondary and postsecondary curricula and their articulation as represented in mid-twentieth century initiatives like Bloom's Taxonomy, as already noted.

Given that the *Entwurf* does not recommend philosophy in and of itself as appropriate to the developmental age of high school students, it is interesting that the most famous of the textbooks that disseminate official Herbartianism was a textbook written for a course on the secondary level preparatory for the systematic study of philosophy proper at the University level. One of Franz Exner's students in Prague, Robert von Zimmermann (1824–1898), authored the textbook that was arguably the most central to Austrian secondary and post-secondary education in the latter nineteenth century, the *Philosophische Propädeutik* (Zimmermann 1860),[19] which provides us with precise details about how Austrian Herbartian classrooms were intended to be implemented. Zimmermann had studied both at the Charles University and at Vienna, combining philosophy with mathematics and natural sciences.[20] His post-PhD career was solid: in 1847, he became an *Assistant* at the University Observatory in Vienna, then was called as an *außerordentlicher Professor* to Olmütz. In 1852, he returned to the Charles

19 The education ministry commissioned him to do this *Philosophical Propaedeutic* in 1851; he completed the first edition in 1852, then did an introduction to a second edition in 1860 and a third edition in 1867, which was translated into several other languages. See Arens 2016.

20 Coen 2007 emphasizes that Zimmermann was a student of Franz S. Exner (1802–1853), following his reinterpretation of judgment as related to probability (p. 61).

University as an *Ordentlicher Professor* of philosophy. In 1861, he was called to Vienna, where he worked until retirement, having also been named a member of the *Kaiserliche Akademie der Wissenschaften* as of 1869.

Zimmermann is unquestionably a *Herbartianer*, both in this training and by self-identification. Zimmerman saw Herbart as part of a specific narrative about how philosophy developed in the modern era, most visibly in *Leibnitz' Monadologie: Deutsch mit einer Abhandlung über Leibnitz' und Herbart's Theorien des wirklichen Geschehens* (1847). In the introduction to the 1860 second edition of his *Propädeutik*, Zimmermann defines his text as providing a preparation for the systematic study of logic at university, in a curriculum that had been expanded in classroom hours a second time, after the 1856 expansion of the hours devoted to this propaedeutic class. Moreover, not only had the curriculum hours expanded, he found that there were too few experienced teachers available, and so he made his second version much clearer as to exact order in classrooms.

The axis for Zimmermann's detailed presentation that combines a textbook with teacher training is Herbart's "excellent distinction between a 'concept in the psychological sense' and a 'concept in a logical sense'" (Herbart's "hervorgehobene[r] Unterschied zwischen 'Begriff im psychologischen' und 'Begriff im logischen Sinne'" [Zimmermann 1860, p. vii]). The logical sense of a concept is objective, which the psychological sense can transform, and it is the goal of his presentation to show how the two relate in creating site-specific and historically distinct forms of knowledge:

> [...] das Verhältniss der *an sich giltigen Wahrheit*, und des zur Erkenntniss wandelbar strebenden *psychischen Denkens* in's Klare gesetzt werden. Dasselbe bleibt das Alpha und Omega echter Philosophie, wenn sie statt subjectiver *Willkür allgemeingiltige* Erkenntniss zu werden sucht.[21]

That difference would have echoed Bolzano's distinction between *Begriff* and *Vorstellung*, with the latter being psychological representation, and the former, a concept clarified for use. Together, this distinction grounds Zimmermann's "Theorie der Induktion" that must be at the core of the sciences in all their forms, not just philosophy (Zimmermann 1860, p. viii). He knows, therefore, that his "introduction to philosophy" is part of a general critique of scientific reasoning as applied to specific realms of human activity.

The *Prolegomena* to the *Propädeutik* outlines how this induction should be taught systematically, starting with perception ("Alles menschliche Wissen kommt zu Stande entweder durch ein *Wahrnehmen* oder durch ein *Nachdenken*

21 Zimmermann 1860, p. vii.

über das Wahrgenommene" [Zimmermann 1860, p. 1]), with the physical senses and the classificatory problems that motivate the student to resolve conflicts between what is known and what is sensed, and hopefully to give up personal opinions in order to reflect on its *Giltigkeit* [sic – a phonetic spelling for *Gültigkeit*], on the limits and uses of any perceived facts used as concepts. That act of reflection leads beyond opinion to knowledge, the

> Richtigkeit jeder Ansicht, als einer Verknüpfung gewisser Vorstellungen unter einander hängt ab: a) von der Richtigkeit und Giltigkeit dieser Vorstellungen selbst und b) von der Richtigkeit und Giltigkeit ihrer Verknüpfung unter einander. Keines von beiden darf fehlen.[22]

It is worth quoting the exact position he takes for the meaning of the sciences involved:

> Logik und Psychologie verhalten sich zu einander wie *Gesetzbuch* und *Naturgeschichte* unserer Gedanken. Jenes zeigt, wie sie gebildet und verknüft werden *sollen*, wenn ihnen Anspruch auf Richtigkeit und Giltigkeit zugestanden werden soll; diese wie sie gebildet und verknüpft *werden*, es mag ihnen dieser Anspruch zukommen oder nicht. Jene sieht daher von der *Wirklichkeit*, dem *Vorkommen* der Gedanken in irgend einem denkenden Wesen gänzlich ab, sagt zwar, *wenn* gedacht werde, um richtig und giltig zu denken, so dürfe nicht anders gedacht werden, als in ihren Formen, schreibt aber weder vor *dass* man denke, noch behauptet sie, dass gedacht *werde*. Die Psychologie dagegen beschäftigt sich lediglich mit der Thatsache *dass* gedacht wird, und untersucht die Bedingungen, unter welchen Gedanken, richtige oder unrichtige, giltige oder ungiltige gleichviel, und im weiten Sinne alle innern Vorgänge, Gefühle, Begehrungen als Seelenerscheinungen zu Stande kommen. Jene erklärt den *Sachverband*, diese den *Thatbestand* unserer Gedanken.[23]

The *Sachverband*, the framework reflecting habituated connections of facts, refers to the perceptions of contents; the latter, part of the individual spirit (Geist).

And the ultimate result for teaching this way is the ability to create an understanding that can be considered a science, a field of knowledge that produces systematic and reproducible knowledge that conforms with both logic and group experience as best it might be:

> Es gibt daher dreierlei Logiken, die doch wesentlich eins sind: a) die *natürliche* d. i. das richtige Verfahren bei Bildung und Verknüpfung der Begriffe, das nach Regeln ohne Kenntniss derselben verfährt; b) die *wissenschaftliche* d. i. der Inbegriff dieser Regeln selbst; c) die

---

22 Zimmermann 1860, p. 5.
23 Zimmermann 1860, p. 8f.

> *logische Kunst* d.i. die durch Übung gewonnene Fertigkeit nach den gekannten Regeln der Logik zu verfahren.[24]

This is what becomes what has been called Austria-Hungary's characteristic turn of mind: a critique of language. To call it that, however, denies that what is being described here is a much deeper commitment to education as an embodied critical epistemology. This Herbartianism is interested in creating a subject of knowledge who is equipped to choose on the basis of both individual experience and concepts internalized from the community.

This is the Herbartian strategy for education that also led to Ernst Mach's great physics textbooks (*Mechanics*, *Optics*, and *Thermodynamics*), each of which is "historical-critical": it starts by introducing the historically conditioned points of view underlying the concepts in each science in terms of where they originated and what purposes they serve as the epistemological limits acculturated into a science by available technology or norms. His *Analysis of Sensations* (1886) starts with the famous drawing of the observer looking past his nose toward his shoes and into the room: it encourages critical engrossment with the validity and limits of observation. Mach's psychophysics thus needs to be reassessed as a cultural as well as a psychological science, and his writings as his propaedeutic for the hard sciences. In a similar vein, the legacies of the Brentano School as passed on by his students like Anton Marty, and through him to names like Karl Bühler, Kazimierz Twardowski, and Edmund Husserl, fall into this paradigm for a critical epistemology because of the cognitive style that I have traced here as inherent in Austro-Hungarian Herbartianism: a paradigm for knowledge production through the agency of individuals but only within frames of reference comprised within historically situated and embodied communities of knowledge.

References

Aichner, Christof/Mazohl, Brigitte (eds.) (2017), *The Thun-Hohenstein University Reforms 1849–1860. Conception – Implementation – Aftermath* (= Austrian Science Fund [FWF]: PUB 397-G28), Vienna: Böhlau Verlag.

Alexander, Larry/Moore, Michael (2016), "Deontological Ethics", *The Stanford Encyclopedia of Philosophy* (Winter 2016 Edition), Edward N. Zalta (ed.), https://plato.stanford.edu/archives/win2016/entries/ethics-deontological/, visited on April 12, 2021.

Arens, Katherine (2016), "Rereading Herder as Heritor of Idealism: Robert von Zimmermann's *Aesthetics*", in: *Herder Jahrbuch / Herder Yearbook* XIII (November 2016), p. 129–146.

24 Zimmermann 1860, p. 16.

Austria: Ministerium für Cultus und Unterricht (1849), *Entwurf der Organisation der Gymnasien und Realschulen in Oesterreich*, Vienna: Kaiserlich-königliche Hofdruckerei.

Bloom, Benjamin (ed.) (1956), with Max D. Englehart, Edward J. Furst, Walker H. Hill, and David R. Krathwohl, *Taxonomy of Educational Objectives Handbook I: The Cognitive Domain*, New York: David McKay / London: Longman.

Blyth, Alan (1981), "From Individuality to Character: The Herbartian Sociology Applied to Education", in: *British Journal of Educational Studies*, 29, no. 1, p. 69–79. doi:10.2307/3120425.

Coen, Deborah R. (2007), *Vienna in the Age of Uncertainty. Science, Liberalism, and Private Life*, Chicago: University of Chicago Press.

Coriand, Rotraud/Winkler, Michael (eds.) (1998), *Der Herbartianismus. Die vergessene Wissenschaftsgeschichte*, Weinheim: Deutscher Studien Verlag.

Cruikshank, Kathleen (1998), "Der Einfluss der Herbartianer auf die Lehrerausbildung in den USA, 1880–1920", in: Coriand/Winkler 1998, p. 99–108.

Deręgowski, J. B. (2017), "Gustav Jahoda. A Life", in: *Journal of Cross-Cultural Psychology* 48, no. 4, p. 455–460. DOI: 10.1177/0022022117703475.

Hayward, Frank Herbert (1907), *The Secret of Herbart. An Essay on the Science of Education*, London: Watts.

Herbart, Johann Friedrich (1802), *Pestalozzi's Idee eines ABC der Anschauung als ein Cyklus von Vorübungen in Auffassen der Gestalten wissenschaftlich ausgeführt*, Göttingen: Johann Friedrich Röwer.

Herbart, Johann Friedrich (1816), *Lehrbuch zur Psychologie*, Königsberg und Leipzig: August Wilhelm Unzer.

Herbart, Johann Friedrich (1824/1825), *Psychologie als Wissenschaft, neu gegründet auf Erfahrung, Metaphysik und Mathematik, Erster, synthetischer Theil* & 1825 *Zweyter, analytischer Theil*, Königsberg: August Wilhelm Unzer.

Herbart, Johann Friedrich (1888), *Johann Friedrich Herbarts Sämtliche Werke*, ed. by Karl Kehrbach, 3. Band, Langensalza: Hermann Beyer & Söhne, 1888.

Herbart, Johann Friedrich (1894), *Pädagogik aus dem Zweck der Erziehung abgeleitet*, Leipzig: Siegismund & Volkening.

Herbart, Johann Friedrich (1896), *Johann Friedrich Herbarts Pädagogische Schriften*, ed. by Fr. Bartholomäi, 6. Auflage, Langensalza: Hermann Beyer & Söhne.

Herbart, Johann Friedrich (1982), *Johann Friedrich Herbart: Pädagogische Schriften*, 1. Band: *Kleinere Pädagogische Schriften*. ed. by Walter Asmus, Stuttgart: Klett/Cotta.

Jäger, Georg (1982), "Die Herbartianische Ästhetik — ein österreichischer Weg in die Moderne", in: H. Zeman (ed.), *Die österreichische Literatur. Ihr Profil im 19. Jahrhundert (1830–1880)*, Graz: Akademische Druck- und Verlagsanstalt, p. 195–219.

Jahoda, Gustav (2006), "Johann Friedrich Herbart. *Urvater* of Social Psychology", in: *History of the Human Sciences* 19, no. 1, p. 19–38, https://doi.org/10.1177/0952695106062145, visited on April 12, 2021.

Kim, Alan (2015), "Johann Friedrich Herbart", *The Stanford Encyclopedia of Philosophy* (Winter 2015 Edition), Edward N. Zalta (ed.), https://plato.stanford.edu/archives/win2015/entries/johann-herbart/, visited on April 12, 2021.

Maigné, Carole (2017), *Une science autrichienne de la forme. Robert Zimmermann (1824–1898)*, Paris: Vrin.

Maigné, Carole/Trautmann-Waller, Céline (eds.) (2009), *Formalismes esthétiques et heritage Herbartian. Vienne, Prague, Moscou*, Hildesheim, Zürich, New York: Georg Olms Verlag.

Müller, Emil (2000), *Das Paradigma des Herbartianismus unter problemgeschichtlichem Aspekt*, PhD dissertation, Erziehungswissenschaftliche Fakultät der Pädagogischen Hochschule Erfurt, https://web.achive.org/web/20070312104905/http://www.uni-erfurt.de/archiv/2001/0003/mueller_dtd.html, visited on April 12, 2021.

Somr, Miroslav/Hrušková, Lenka (2014), "Herbart's Philosophy of Pedagogy and Educational Teaching (The Views and Differences of Opinion)", in: *Studia Edukacyjne*, no. 33 [Poznań], p. 413–429.

Zimmermann, Robert von (1860), *Philosophische Propädeutik*, 2nd ed., Vienna: Braumüller, https:// archive.org/ details/philosophischep00zimmgoog, visited on April 12, 2021.

Josef Zumr

Der Herbartianismus in Böhmen

Zusammenfassung: Der Herbartianismus gehört zu den einflussreichen Strömungen des theoretischen Denkens in den böhmischen Ländern des neunzehnten Jahrhunderts. Das betrifft vor allem philosophische Propädeutik, Metaphysik, Pädagogik und Ästhetik. Der Rekonstruktion dieses Einflusses und damit verbunden der Offenlegung versteckter Zusammenhänge ist dieser Aufsatz gewidmet.

1 Der Herbartianismus in Böhmen vor 1848

1.1 Franz Exner und die Prager Universität

Der Herbartianismus hat gerade in Böhmen und vornehmlich an der Prager Universität Fuß gefasst und sich von da in den anderen Ländern der österreichischen Monarchie ausgebreitet. Zweifellos war Franz Exner der erste bedeutende Herbartianer an der Prager Universität.

Exner war aber nicht der erste, der sich in Österreich zu Herbart bekannte. Dies war der Wiener Philosophieprofessor Leopold Rembold (1786–1842). Rembold war anfangs ein Anhänger der Gefühlsphilosophie von Jacobi, die von Bayern durch Vermittlung von Jakob Salat ihren Ausgang nahm. Dann nahm Rembold insbesondere die Herbartsche Psychologie gefangen, die ihn zum Herbartianismus führte. Rembold wurde aber schon im Jahre 1824 wegen Freidenkerei seiner Professur an der Universität enthoben und arbeitete dann als praktischer Arzt. Dabei ließ er den Kontakt mit der Philosophie nie abreißen und sorgte auch weiterhin für die Verbreitung seiner Anschauungen in seinem Freundeskreis. Einer seiner Schüler noch während seines Wirkens an der Universität war Exner. Exner erwarb 1825 das Doktorat und trug mehrere Jahre lang an der Wiener Universität Philosophie und Pädagogik vor, bis er 1832 als ordentlicher Professor nach Prag berufen wurde – an jene Stelle, die durch den Abgang von Lichtenfels, ebenfalls ein Schüler Rembolds, der aber der Philosophie Jacobis treu blieb, freigeworden war.

Josef Zumr, Philosophisches Institut der Tschechischen Akademie der Wissenschaften
zumr@flu.cas.cz

https://doi.org/10.1515/9783110747324-006

Franz Exner (1802–1853) erfreute sich bald nach seiner Ankunft in Prag großer Beliebtheit bei seinen Studenten und bei den Prager Intellektuellen. Er brachte einen neuen Geist und neue Gedanken ins Universitätsleben ein, und bei seinen Erklärungen hielt er sich nicht an die vorgeschriebenen Lehrbücher (in Österreich war es bis 1848 Vorschrift, genau nach genehmigten Lehrbüchern zu unterrichten; Exner wurde wegen der Verletzung dieser Vorschrift zur Verantwortung gezogen). Außerdem war er ein ausgezeichneter Pädagoge und verstand es, seine Studenten für seinen Gegenstand zu begeistern. Seine Vorlesungen besuchten tschechische und deutsche Hörer gemeinsam, viele von ihnen traten später als bedeutende Repräsentanten der beiden Kulturen hervor (z. B. der Dichter Karel Hynek Mácha, der Schriftsteller, Kritiker und Politiker Karel Sabina, die Politiker František Ladislav Rieger und Karel Havlíček Borovský, die Philosophen Augustin Smetana, Ignác Jan Hanuš, František Čupr, Gustav Adolf Lindner, die deutschen Philosophen Robert Zimmermann, Joseph Wilhelm Nahlowsky, der Psychologe Wilhelm Volkmann, der Historiker Anton Heinrich Springer u. a.), und alle erinnerten sich ausnahmslos dankbar an ihren Lehrer.[1] An den freundschaftlichen Treffen in Exners Haus nahmen die herausragenden Repräsentanten beider Nationalitäten teil. Stellvertretend für alle seien hier nur der Historiker František Palacky und der Philosoph Bernard Bolzano genannt.

Exners eigene wissenschaftliche Tätigkeit in dieser Zeit ist nicht unerheblich. Er publizierte eine Reihe kleinerer Arbeiten, teils historische, teils polemische, unter anderem: *Über den Nominalismus und Realismus* (1842), *Über Leibnizens Versuch eine allgemeine Wissenschaft des Beurteilens und Erfindens aufzustellen* (1843), *Über Leibnizens Universalwissenschaft* (1845), *Über die Lehre von der Einheit des Denkens und Seins* (1848). Wesentlich bekannter wurden jedoch seine zwei psychologischen Abhandlungen in zwei Bänden: *Die Psychologie der Hegelschen Schule* (Leipzig 1841, 1842), und die Rezension des im Jacobischen Sinne verfassten *Lehrbuches der Psychologie* von J. R. Jäger (1841), zu letzterer bemerkte Robert Zimmermann später, dass Exner

> der Jacobischen Philosophie den Absagebrief schrieb und der bis dahin einflussreich gewesenen Psychologie Jacobis für immer ein Ende machte, wie er es [...] später durch seine in Witz und Scharfsinn unvergleichliche Beurteilung der Psychologie der Hegelschen Schule der noch einflussreicheren Psychologie Hegels bereitet hat.[2]

Dauernde Verdienste erwarb Exner jedoch als Reformator der österreichischen Mittelschulen und Universitäten. Reformgedanken hegte er bereits in den dreißiger

---

1 Siehe Jánský 1958.
2 Zimmermann 1888, p. 265.

Jahren (*Die Stellung der Studierenden auf den Universitäten* und *Über die Bedeutung des akademischen Studiums*, 1837) und er war durch seine Kenntnis der Herbartschen Pädagogik theoretisch gut vorbereitet. In den Jahren 1845–1847 wirkte er in Wien als Mitglied der Kommission für Schulreform. Im Jahre 1848 wurde er wieder nach Wien berufen und zum Ministerialrat ernannt. In der *Wiener Zeitung* veröffentlichte er den „Entwurf der Grundzüge des öffentlichen Unterrichtswesens in Österreich",[3] dessen programmatische Grundsätze dann vor allem durch Hinzutun des Ministers Graf Leo Thun-Hohenstein verwirklicht wurden. Exner ging es um die Errichtung eines Gymnasiums mit acht Klassen und die Reform des Universitätsstudiums, bei dem die maximale Freiheit der Wissenschaft garantiert werden sollte. Diesem Bestreben war nach dem Antritt der Reaktion in der Zeit nach der Revolution jedoch kein Erfolg beschieden, wodurch die österreichischen Universitäten bis in die siebziger Jahre unter einer gewissen ideologischen Aufsicht der katholischen Kirche blieben. Aber: „in der Situation von 1848 war Exner sicher als Liberaler zu bezeichnen, er war ein Anhänger der voraussetzungslosen Wissenschaft, wie sie sich an den deutschen Universitäten herausgebildet hatte".[4] Große Hilfestellung bei der Erarbeitung des Reformprogramms erhielt Exner von einem anderen Herbartianer, dem Professor für klassische Philologie am Stettiner Mariengymnasium, Hermann Bonitz (1814–1888), den Exner schon früher in Berlin kennengelernt hatte und der auf Exners Empfehlung hin 1849 ebenfalls nach Wien berufen wurde. Durch die Zusammenarbeit der beiden Reformatoren entstand ein neuer *Entwurf der Organisation der Gymnasien und Realschulen in Österreich*. „Er war von dem Ideal der allseitigen Bildung bestimmt und bemühte sich, neben den klassischen Sprachen auch die Mathematik und die Naturwissenschaften ausreichend zu berücksichtigen."[5] Von seiner durchdachten Qualität zeugt die Tatsache, dass er sich als Modell noch tief bis ins 20. Jahrhundert hielt.

1.2 Augustin Smetana

Zu Beginn müssen wir zwei Vorbehalte gegenüber diesem Unterkapitel formulieren.

Erstens: Gegenstand unserer Arbeit ist die Skizzierung der Geschichte der deutschen und österreichischen Philosophie in Böhmen. Augustin Smetana war

3 Vgl. Zimmermann 1848.
4 Lentze 1962, p. 30.
5 Lentze 1962, p. 31.

ein selbstbewusster Tscheche und gehört deshalb nicht zur deutschen Philosophie. Die Antwort lautet, dass er zu einem gewissen Grad doch dazu gehört. Smetana schrieb seine Bücher auf Deutsch, setzte sich darin mit der deutschen Philosophie auseinander und kommunizierte damals mit der deutschsprachigen Welt. Er war kein engstirniger Nationalist, und wenn er von seiner tschechischen Heimat sprach, so geschah dies immer im allgemein menschlichen Kontext. Deshalb erachten wir es für berechtigt, ihn im Zusammenhang mit der deutschen Philosophie zu behandeln. Aus denselben Gründen machen wir bei Otakar Hostinský eine Ausnahme. Im Gegensatz dazu streifen wir vollständigkeitshalber jene Herbartianer, die vor allem tschechisch schrieben und mit der deutsch-österreichischen Philosophie nur in losem Zusammenhang stehen, nur flüchtig.

Zweitens: Smetana wird üblicherweise als Hegelianer bezeichnet. Dem kann man zustimmen. Trotzdem war Smetana im internationalen Vergleich einer der wenigen Philosophen, die auf kreative Weise in ihr System auch Herbart einbauten. Mit Recht gebührt ihm somit ein Platz in der Geschichte des Herbartianismus.

Augustin Smetana (1814–1851) war Priester des Ordens der Kreuzherren mit dem roten Stern. Er studierte Theologie und zugleich Philosophie bei Fr. Exner, wirkte von 1842–1845 als Exners Adjunkt, 1842 unternahm er eine Reise durch Deutschland, wo er sich mit der Hegelschen Philosophie vertraut machte, und 1846–1847 wurde er statt des abwesenden Prof. Exner mit Philosophievorlesungen beauftragt. 1848 wird er zum Dekan der Philosophischen Fakultät gewählt, beteiligt sich aktiv an der Revolution, trägt noch 1849 vor, aber Ende des Jahres verliert er seine Stelle an der Universität, im Frühjahr 1850 tritt er öffentlich aus der Kirche aus, wird nachträglich exkommuniziert und ist bis zu seinem vorzeitigen Tod dem Druck seitens der Kirchenbehörden ausgesetzt.

Smetana war der einzige tschechische Denker, der im Geiste des Links-Hegelianismus den Sinn der Revolution 1848 philosophisch durchdachte und versuchte, eine entsprechende geschichtsphilosophische Konzeption zu schaffen. Er glaubte, in der Revolution den Beginn der humanistischen Epoche zu sehen, die in eine sozialistische Gesellschaft münden würde. Zum Verständnis dieses Geschichtsprozesses dienten Smetana als passendstes Mittel die Geschichte der Philosophie sowie die Lehren daraus.

Die Geschichte der Philosophie ist nach Smetanas Meinung seit ihrem historischen Beginn eine ständige Bemühung um die Erreichung der Identität von Wissen und Sein.[6] Diese Bestrebung bildet den Anfang und das Ziel des ge-

6 Vgl. Smetana 1850, p. 2.

samten Philosophierens, d. h. des Prozesses, der der wesentlichste Ausdruck des Menschentums ist, ein Ausdruck des Suchens nach dem Sinn der menschlichen Existenz und der Existenz der Gesellschaft überhaupt. Durch die Erreichung dieser Einheit von Wissen und Sein wird der Charakter des menschlichen Bewusstseins erhellt, ohne sie würde die gesamte Philosophie ihre eigene *raison d'etre* verlieren. In der Geschichte der Philosophie geht es nach Smetana um die rationale Erläuterung der Identität von Wissen und Sein, denn diese Identität ist eine ontologische Tatsache, nämlich das Wesen des menschlichen Bewusstseins.

Diese Voraussetzung stützt Smetana auf die historische Erfahrung der Menschheit, wonach der Mensch aus dem göttlichen Geist und dem irdischen Körper besteht. Diese Elemente, d. h. Wissen und Sein, verschmelzen in einem bestimmten Augenblick im menschlichen Bewusstsein zu einer Einheit, und daher können ihre Attribute, das Göttliche und das Irdische oder das Unendliche und das Endliche, wie sie Smetana auch nennt, dialektisch ihre Stelle wechseln und sich mit ihrem Gegensatz verbinden. Das göttliche Attribut gehört nicht nur dem Geiste: das Sein, das durch das Wissen im Bewusstsein zum Selbstbewusstwerden gelangt, nimmt auch den Charakter des Unendlichen an, und umgekehrt ist das Wissen, das sich ändert, vergänglich. Im Zusammenhang damit führt Smetana eine gewisse Typologie der bisherigen Lösungen dieser Frage durch und teilt die gesamte Philosophie in einen östlichen Typ, den die Betonung des göttlichen Seins charakterisiert, und in einen westlichen, der das irdische Wissen betont. In beiden Typen sind aber die einzelnen Seiten nicht so weit verabsolutiert, dass sie ihr Gegenteil aufheben würden.[7] Die Lösung des Verhältnisses zwischen Sein und Wissen durchlief in der geschichtlichen Entwicklung des philosophischen Denkens viele Modifikationen, und jede Epoche steuerte ihren Anteil dazu bei, so dass der gesamte Prozess eine emporstrebende Tendenz aufweist. Smetana unterscheidet drei historische Etappen dieses Prozesses, in denen sich das menschliche Bewusstsein entwickelt.[8] Die abschließende Phase der zweiten Etappe, die deutsche Philosophie von Kant an, stand – nach der Meinung Smetanas – vor der endgültigen Lösung des gesamten Problems, denn in den Systemen von Hegel und Herbart synthetisierte sie die gesamte östliche und westliche Weisheit und näherte sich, wie niemals zuvor, der Erneuerung der Einheit von Wissen und Sein auf einer höheren Stufe. Durch eine richtige Vermittlung zwischen der Philosophie von Hegel und Herbart könne daher das Endziel erreicht werden. Das ist die Aufgabe der dritten Etappe und diese Aufgabe

7 Siehe Smetana 1850, p. 15.
8 Siehe Smetana 1850, p. 34f.

erfüllt die ganze Denkbestrebung Smetanas, der mit allen seinen Kräften zur Erreichung dieses Zieles beitragen wollte.

Smetanas Historismus, belehrt von Hegel und den ersten hegelianischen Philosophiehistorikern, bildet ein weitgehend selbständiges und von scharfsinnigen dialektischen Beobachtungen durchdrungenes Bild der philosophischen Vergangenheit. In der bemerkenswerten Entwicklung des deutschen Idealismus sieht er völlig richtig eine gewisse Reproduktion des vorhergegangenen Stadiums des europäischen Denkens: Kant und Fichte knüpfen an Descartes an, Schelling und Hegel an Spinoza; Herbart ist eine gewisse Modifikation von Leibniz. Aber während Leibniz gewissermaßen ein Vermittler zwischen Descartes' Szientismus und Spinozas Substantialismus war, oder mit den Worten Smetanas, „zwischen dem abendländischen und morgenländischen Element", unterschiedet sich Herbart deutlich vom ganzen Zyklus von Kant bis Hegel als ausgeprägter Vertreter der Philosophie des Wissens.[9] Und zwar deshalb, weil Herbart das substantielle Element der Leibnizschen Monadologie ablehnte, das allerdings bei Leibniz der Träger der Entwicklung, der Historizität war, und weil er das Prinzip des Wissens in seiner reinen Form annahm. Das Positive dieser Einstellung war die konsequente Ablehnung jedweder angeborener Ideen, und daher eine weitere Emanzipation der Philosophie von theologischen Elementen. Dadurch unterstrich Herbart auch seine Verschiedenheit vom Spinozismus, vor allem an Hegel und Schelling denkend.[10] Smetana stellt richtig fest, dass Herbarts Bemühung um eine Differenzierung zwar seinen Widerspruch zu Hegel steigert, jedoch kann er die gemeinsamen Züge beider Systeme nicht leugnen.[11] Diese sind bei Herbart als idealistischer Realismus und bei Hegel als realistischer Idealismus charakterisiert. Der gegenseitige Zusammenhang beider Denker (und auch der übrigen Vertreter des deutschen Idealismus) ist dadurch gegeben, dass sie alle um eine positive Lösung der Widersprüche innerhalb des Bewusstseins (zwischen Wissen und Sein) bemüht waren, wenn sie auch einseitig immer nur eine bestimmte Seite unterstrichen. Hegels Beitrag zur Lösung dieser Frage sieht

9 „Herbart ist der Gipfelpunkt der gesamten occidentalischen Weltweisheit als solcher." (Smetana 1850, p. 271).

10 In seiner handschriftlichen Bemerkungen über Herbart machte Smetana auf die Tatsache aufmerksam, dass Herbarts ständige Bemühung, sich vom Hauptstrom des deutschen Idealismus zu distanzieren, bewirkt, dass viele eigenwillige Elemente in sein System eindringen, die das System entwerten und die nur durch diese Opposition erklärlich sind (siehe Literarisches Archiv – Literární archiv PNP, Prag, VII, Nr. 58, XII, p. 7).

11 „Wir möchten von Hegel und Herbart nicht mehr, wie von Spinoza und Leibniz, sagen, dass sie auf demselben Boden stehen; wohl aber, dass ihre Systeme denselben, aus den Elementen des Realismus und Idealismus zusammengesetzten Geist atmen." (Smetana 1850, p. 271f.).

Smetana darin, dass er zum Unterschied von Spinoza – der zur Übereinstimmung der Attribute des Denkens und der Ausdehnung noch ein weiteres Prinzip, nämlich die unbewegliche göttliche Substanz forderte – dieses Verhältnis durch einen dialektischen Prozess des Geistes selbst festzuhalten wusste. Smetana stimmt mit Hegel darin nicht überein, dass dieser die Identität zwischen Sein und Wissen für absolut hält, und sieht hingegen das Verdienst Herbarts im Beweis, dass es sich um eine relative Identität handelt. Diese Feststellung ist für Smetana gerade vom Gesichtspunkt ihrer gesellschaftlichen Folgen von außerordentlicher Bedeutung.

Verlassen wir aber für einen Augenblick Smetanas Terminologie und betrachten wir das ganze Problem von einem anderen Gesichtspunkt. Smetana war sich dessen wohl bewusst, dass die ganze deutsche Philosophie der klassischen Epoche eine Polemik mit der traditionellen substantiellen Auffassung des Verhältnisses zwischen Realismus (Materialismus) und Idealismus, mit der metaphysischen Lösung des Verhältnisses von Subjekt und Objekt ist. Es war das Verdienst von Hegel und Herbart, dass der metaphysische Substanzionalismus weitgehend liquidiert wurde. Einerseits wurde er durch die dialektische Bewegung des ewigen Werdens und andererseits durch die funktionelle Beziehung der autonomen Elemente ersetzt. In beiden Fällen handelte es sich um eine weitere Verweltlichung des Denkens und dadurch um einen Fortschritt im Prozess der Selbstbefreiung des Menschen. Warum also kritisierte Smetana Hegel wegen seiner Ansicht über die absolute Identität des Denkens und des Seins und wegen des daraus folgenden einseitigen Objektivismus?[12] Das hängt offensichtlich mit Smetanas Gesamtkonzeption der Zukunft der Philosophie und also auch der Menschheit zusammen. Für das dynamische Bild der Geschichtsphilosophie bis zur Schwelle der Gegenwart konnte Smetana die grundlegenden Charakterzüge von Hegels großartigem Aufriss der Entwicklung des objektiven Geistes ausnützen. Der Sinn dieser Bewegung konnte die Befreiung des menschlichen Geschlechtes, keinesfalls aber die völlige Befreiung des Individuums sein. Wie schon gesagt, verband Smetana Hegels Philosophie mit dem Typus der östlichen Weisheit, für den es charakteristisch ist, dass er das Individuelle dem Allgemeinen und die Freiheit der Notwendigkeit im Namen irgendeines höheren Prinzips völlig unterordnet. Das war jedoch mit Smetanas Vorstellung über die zukünftige Gesellschaft, eine humanistische, sozial gerechte Gesellschaft, ganz unvereinbar. Im Zusammenhang mit der Revolution des Jahres 1848 traten mehr

---

12 Diesen Vorwurf wiederholt Smetana konsequent nicht nur in seiner Schrift *Die Katastrophe und der Ausgang der Geschichte der Philosophie*, sondern auch in seinem wichtigsten systematischen Werk: *Der Geist, sein Entstehen und Vergehen*, 1865.

als je zuvor Ansichten auf, die eine positive Lösung der Freiheit des Individuums im Rahmen der neuentstandenen Gesellschaft forderten. Das waren durchwegs radikale Ansichten, die aus den Traditionen der revolutionären Romantik und des romantischen Titanismus entstanden und oft in den anarchistischen Individualismus übergingen. Smetana jedoch verlangte ein harmonisches Verhältnis zwischen dem Einzelnen und der Gesellschaft. Es ist charakteristisch, dass er die philosophische Begründung dieses edlen Ideals gerade in der Philosophie Herbarts suchte, deren Schöpfer selbst aber zur Revolution ein völlig negatives Verhältnis hatte. Smetana findet diese Begründung in Herbarts Gedanken über die relative Identität von Wissen und Sein. Diese Identität ist hinreichend, in Smetanas Interpretation, für die relative Selbständigkeit des Objekts, sodass sie seine geschichtliche Bewegung sicherstellt und gleichzeitig dem autonomen Subjekt auch einen angemessenen Raum darbietet.[13]

Der Gedanke der relativen Identität des Seins und des Wissens ist die äußerste Grenze, zu der die Lösung der Beziehung zwischen Subjekt und Objekt auf dem Boden der Philosophie, oder genauer gesagt, auf dem Boden des philosophischen Wissens, gelangen kann. In diesem Sinne ist, der Überzeugung Smetanas gemäß, das ganze Problem schon genügend demystifiziert, d. h. von theologischen Überlagerungen befreit, denn es wurde zum Problem des menschlichen Bewusstseins, des Bewusstseins des menschlichen Individuums, seiner Entstehung und Entwicklung.

> Die Differenz zwischen dem Göttlichen und dem Menschlichen, die bei Hegel nur noch eine formelle ist, schwindet gänzlich, wenn ihr gegenüber das Herbartsche Prinzip der Individualität zur vollen Gültigkeit erhoben, wenn überhaupt nicht das Bewusstsein des Absoluten, sondern das des menschlichen Individuums abgeleitet und wenn die relative Identität des Seins und des Wissens im menschlichen Bewusstsein, die unmittelbar vorhanden ist, in ihrem vermittelten Charakter nachgewiesen werden wird.[14]

In solcher Auffassung bedeutet dieses Problem auch die völlige Liquidierung der christlichen Philosophie und dadurch auch der Philosophie im alten Sinne des Wortes überhaupt. Die neue Philosophie des befreiten Bewusstseins (zu deren Geburt Smetana mit seiner Analyse der Geschichte der Erkenntnis beisteuern wollte und als deren grundlegende Aufgabe er die konkrete Feststellung des

---

13 „Wohl kann die Wahrheit des individuellen und die Bedeutung des höheren Lebens nur durch eine solche Auffassung erhalten werden [...]" (Smetana 1850, p. 278).

14 Smetana 1850, p. 284. Es ist interessant, dass Smetana in seiner früheren Schrift *Die Bedeutung des gegenwärtigen Zeitalters* (1848) die Endphase dieses Prozesses mit Feuerbach verbindet.

Schnittpunktes der relativen Identität von Wissen und Sein im Bewusstsein des Individuums bestimmte) nähert sich dem Ziel, zur allgemeinen Harmonie und höheren Einheit alle Widersprüche, in denen das Denken ständig vergebens herumkreiste, kommen zu können, wenn sie zur Praxis wird.[15]

Smetana ist in der Geschichte der modernen tschechischen Philosophie der erste Philosoph, der das philosophische Problem der Praxis in sein System als einen organischen Bestandteil aufnahm. Sein großes Verdienst war die Feststellung, dass die alten philosophischen Mittel zur weiteren Lösung der Beziehung zwischen Subjekt und Objekt nicht mehr ausreichen und dass daher neue Mittel dazu benützt werden müssen, die in den Bereich der Praxis gehören.

Das Ziel, zu dem die neue Wissenschaft, die Philosophie des befreiten Bewusstseins, gelangen soll, ist die Kunst und die Liebe. Die Kunst, die eine Verneinung der Religion ist, wird eine der Religion der alten Gesellschaft entgegengesetzte Funktion haben. Die Religion war eine Kraft, die den Menschen im Namen eines höheren unendlichen Wesens unterwarf und so den Menschen seiner Individualität beraubte. Die Kunst hingegen konzentriert den Menschen auf „göttliche" Kräfte in ihm selbst und bewirkt das Aufblühen seiner Persönlichkeit. Bei Smetana handelt es sich jedoch um die Harmonie, um die relative Identität. Die Liebe, die in der zukünftigen Gesellschaft das Recht ersetzt, ist der zweite Pol der Beziehung, denn sie ist eine Verneinung des Egoismus, ein humanistisches Bindeglied zwischen Individuum und Gesellschaft und eine Negation des Rechtspartikularismus. Die Praxis tritt hier also in der Gestalt zweier menschlichen Aktivitäten auf, die von ihrem Schnittpunkt aus im befreiten Bewusstsein in entgegengesetzte Richtungen weisen.[16]

Das war also die Lehre, welche Smetana aus dem revolutionären Zeitalter zog und die er seinen Zeitgenossen vorgelegt hat.

2 Der Herbartianismus in Böhmen nach 1848

2.1 Robert Zimmermann

Robert Zimmermann (1824–1898) war ein gebürtiger Prager und verbrachte nahezu die Hälfte seines Lebens in Böhmen, wovon er zehn Jahre Professor an der Prager

15 Den Ausdruck „Praxis" benützt Smetana nicht, es handelt sich allerdings um dieses Problem.

16 Die Dialektik der Kunst, der Liebe und der Philosophie des befreiten Bewusstseins entwickelte Smetana in den ersten drei Kapiteln des Buches: *Die Bedeutung des gegenwärtigen Zeitalters als unmittelbares Ziel der revolutionären Zeit*. Das Buch erschien gerade im Jahre 1848.

Universität war. In Prag entstand auch seine *Ästhetik*, sein bemerkenswertestes Werk, durch das er nicht nur in die Geschichte des Herbartianismus einging, sondern in die Geschichte der Philosophie überhaupt. Deshalb fällt sein Wirken in Böhmen in den Rahmen unseres Themas.

Zimmermanns Vater Johann August war Gymnasiallehrer in Iglau (Jihlava), Pisek (Písek) und schließlich in Prag, von wo er 1844 als Mitglied der Hofkommission für die Reform des Mittelschulstudiums nach Wien berufen wurde. Er gehörte zu den engen Freunden von B. Bolzano, in dessen Freundeskreis auch sein Sohn Robert verkehrte, der unter dem Patronat von Bolzano Philosophie, Mathematik und Naturwissenschaften studierte. An der Universität besuchte Zimmermann die philosophischen Vorlesungen des Herbartianers Exner, und der Einfluss von diesen beiden Seiten bestimmte auch seinen gedanklichen Werdegang. 1847 geht er nach Wien und wird Assistent an der dortigen Sternwarte. 1849 habilitiert er sich an der Wiener Universität im Fach Philosophie, 1850 wird er an der Universität in Olmütz zum Professor ernannt, von wo er zwei Jahre später auf die Prager Universität überwechselt. In Prag wirkt er bis 1861 und hält dann bis 1895 an der Wiener Universität Vorlesungen, wo ihm zahlreiche wissenschaftliche Ehrungen zuteil werden (u. a. 1869 Mitglied der Kaiserlichen Akademie der Wissenschaften). Noch während seiner Prager Jahre gibt er 1852 die im Herbartschen-Bolzanoschen Geist konzipierte *Philosophische Propaedeutik* heraus, die mehrere Neuauflagen erfuhr und Jahrzehnte hindurch in ganz Österreich als Mittelschullehrbuch verwendet wurde.[17]

Wie bereits erwähnt, ging es in Zimmermanns wichtigstem Werk seiner Prager Zeit um die Ästhetik, die er immer in enger Verbindung mit der Philosophie sah. Seine Bedeutung für die allgemeine Ästhetik hat zwei Aspekte. Zimmermann war der erste Ästhetiker, der einen systematischen und ausführlichen Überblick über die Geschichte seines Gegenstandes gab. In unserem Kontext dürfen wir auch den Umstand nicht übergehen, dass seine *Geschichte*

17 Cf. Zimmermann (1852a). In der zweiten (1860) und dritten (1867) Auflage dieses Buches wurde Bolzanos Einfluss unterdrückt, was nicht nur mit der engeren Zuneigung an Herbart zusammenhing, sondern auch damit, dass Zimmermann von einigen katholischen Theologen als Anhänger der verurteilten Ansichten Bolzanos denunziert wurde. Siehe dazu bes. Winter 1975. Zimmermann veröffentlichte in dieser Zeit noch folgende Arbeiten: *Leibniz' Monadologie mit einer Abhandlung über Leibnizens und Herbarts Theorien des Wirklichen Geschehens* (1847); *Über den wissenschaftlichen Charakter und die philosophische Bedeutung Bernard Bolzanos* (1849); *Leibniz und Herbart, eine Vergleichung ihrer Monadologien* (1849); *Was erwarten wir von der Philosophie?* (1852); *Das Rechtsprinzip bei Leibniz* (1852); *Über das Tragische und die Tragödie* (1856); *Schiller als Denker* (1859).

der Ästhetik als philosophischer Wissenschaft[18] auf dem Boden der Prager Universität entstand. Zimmermanns Geschichte ist mit ausgezeichneter Erudition geschrieben und verfolgt die gegebene Problematik von der Antike bis zur Gegenwart. Er beschreibt die Entwicklung des ästhetischen Denkens äußerst objektiv und sachlich, ohne in den akademischen Objektivismus zu verfallen. Zimmermann ist von seinem System überzeugt und bringt aus seiner Sicht eine begründete Kritik der Ansichten seiner Vorgänger. Erwähnenswert ist auch, dass Zimmermann hier zum ersten Mal die ästhetischen Anschauungen seines Lehrers B. Bolzano in die historischen Zusammenhänge aufnahm und sie zumindest vorübergehend davor bewahrte, völlig in Vergessenheit zu geraten.

Ein weiteres Verdienst Zimmermanns war, dass er in seinem Buch *Allgemeine Ästhetik als Formwissenschaft*[19] erstmals ein konsistentes System der Herbartschen Beziehungs- oder Formästhetik ausgearbeitet hat. Herbart selbst schrieb bekannterweise keine systematische Ästhetik, obzwar er zahlreiche grundlegende Thesen dazu aufstellte. Er verfasste aber eine sog. praktische Philosophie, d.h. eine Ethik, die seiner Auffassung nach als Lehre über die Willensverhältnisse ein eigenes Gebiet der allgemeinen ästhetischen Beziehungen war. Zimmermann kehrte dieses Verhältnis in gewissem Sinne um und schuf nach dem Herbartschen System der Ethik sein System der allgemeinen Ästhetik, die seiner Meinung nach eine normative Wissenschaft ist. Wenn auch diese Vorgangsweise später – insbesondere von Hostinský – kritisiert wurde, so wird dadurch jedoch nicht der Wert aller Bestrebungen Zimmermanns in Zweifel gezogen. Er arbeitete in vielen Richtungen ganz selbständig und konfrontierte seine Anschauungen auch mit der damaligen künstlerischen Wirklichkeit. Tatsache bleibt, dass er der damals schon zerbröckelnden idealistischen Ästhetik ein festes Begriffssystem entgegenstellte, das viel tiefer als die bisherigen Theorien zum Wesen des Ästhetischen durchdringen konnte und das Voraussetzungen für die weitere Entwicklung des ästhetischen Denkens schuf. Außerdem trägt die Ästhetik in seiner Auffassung deutlich humanistische Züge, in denen Zimmermann einerseits seinen Lehrer Bolzano nicht verleugnen kann und in denen sich andererseits die Inspiration Schillers zeigt, die bereits dem ursprünglichen Projekt Herbarts eigen war.

Zimmermann gehörte an der Prager Universität zu den populären Professoren; großer Beliebtheit erfreute er sich bei seinen tschechischen Hörern. Zu seinen Prager Schülern gehörten auch die später bedeutenden Repräsentanten des tschechischen Herbartianismus Josef Dastich und Josef Durdík, bei denen das Schüler- offenkundig zu einem Freundschaftsverhältnis wurde. Eine bemerkens-

---

18 Zimmermann 1858.
19 Zimmermann 1865.

werte Persönlichkeit blieb Zimmermann auch während seines Wirkens an der Wiener Universität.[20] Einer seiner Schüler in Wien war auch Tomáš G. Masaryk.

Zimmermann war in der zweiten Hälfte des 19. Jahrhunderts nicht der einzige Repräsentant des deutschen Zweiges des Herbartianismus in Böhmen. Außer ihm und nach ihm wirkten hier weitere bedeutende Herbartianer. Einige von ihnen hielten auch Philosophievorlesungen, aber ihr wissenschaftlicher Beitrag hatte in anderen Fächern seinen Schwerpunkt. Vollständigkeitshalber führen wir hier wenigstens ihre Namen an. Es war vor allem der aus Prag stammende Wilhelm Volkmann (1821–1877), der dreißig Jahre lang an der Prager Universität Philosophie und Ästhetik vortrug; die Hauptdisziplin, der er sich am meisten widmete, war jedoch die Psychologie. Neben zahlreichen anderen Werken war er der Autor des zweibändigen Werks: *Grundriss der Psychologie vom Standpunkte des philosophischen Realismus und nach genetischer Methode* (1856); dieses Werk erschien in mehreren Auflagen und in Übersetzung in mehrere Sprachen. Kurz wirkte in den böhmischen Ländern (Prag, Olmütz) auch Josef Wilhelm Nahlowsky (1812–1885) – noch einer der ersten Schüler Exners – bevor er Professor in Graz wurde. Lose in die Reihe der Herbartianer aufnehmen kann man auch Otto Willmann (1839–1920), ab 1872 Professor in Prag, der philosophisch eher zur Geschichte der katholischen Philosophie gehört, in der Pädagogik sich aber zu Herbart bekannte und Herausgeber dessen pädagogischer Arbeiten wurde.[21]

2.2 Die tschechischen Herbartianer

Wenn wir im vorangehenden Kapitel festgestellt haben, dass wir im Überblick über die deutsche und österreichische Philosophie in Böhmen vollständigkeitshalber auch die tschechischen Herbartianer erwähnen wollen, so war die Vollständigkeit nicht der einzige Grund. Außer Augustin Smetana und Otakar Hostinský, denen wir aufgrund ihrer außerordentlichen Bedeutung besondere

20 Ein interessantes Zeugnis davon findet man in Erinnerungen des berühmten Theosophen Rudolf Steiner, der zum Ende der siebziger Jahre Hörer bei Zimmermann an der Universität war: „Robert Zimmermann war eine merkwürdige Persönlichkeit. Er hatte eine ganz ungewöhnlich hohe Stirn und einen langen Philosophenbart. Alles an ihm war gemessen, stilisiert. Wenn er zur Türe hereinkam, aufs Katheder stieg, waren seine Schritte wie einstudiert und doch wieder so, dass man sich sagte: dem Mann ist es selbstverständlich-natürlich, so zu sein. Er war in Haltung und Bewegung, wie wenn er sich selbst dazu nach Herbart'schen ästhetischen Prinzipien in langer Disziplin geformt hätte. Und man konnte doch rechte Sympathie mit alledem haben." (Steiner 1925, p. 35).

21 Für ausführliche Angaben über alle diese Herbartianer siehe insbesondere Tretera 1989.

Aufmerksamkeit schenken, publizierten alle tschechischen Herbartianer ihre Arbeiten teilweise auf Deutsch. Diese Arbeiten stießen auch im deutschsprachigen Gebiet auf ein gewisses Echo, die tschechischen Denker waren mit ihren deutschen Kollegen meist durch freundschaftliche Kontakte verbunden, und gemeinsam schufen sie in Prag und Böhmen eine gewisse intellektuelle Atmosphäre, die oft auf andere österreichischen Länder übergriff.

Zur ersten Generation der tschechischen Herbartianer, die noch Schüler von Exner waren, gehört František Čupr (1821–1882). Die Herbartsche Philosophie verteidigte er bereits im Jahre 1847 in der Diskussion unter den tschechischen Intellektuellen darüber, ob eine nationale Kultur überhaupt eine Philosophie braucht und ob es nicht schädlich ist, sich mit der deutschen Philosophie auseinanderzusetzen. Zusammen mit Augustin Smetana wehrte er die nihilistischen und nationalistischen Ausfälle ab und verteidigte auf überzeugende Weise die Nützlichkeit der Philosophie. Wegen seiner politisch nonkonformen Haltungen wurde ihm jedoch eine akademische Laufbahn unmöglich gemacht. Später widmete er sich der Religionsphilosophie („altindische Lehre"), wo er manche nicht ganz authentisch interpretierten Thesen Herbarts mit Elementen von Schopenhauer ergänzte.

Von Exners tschechischen Schülern verdient noch Gustav Adolf Lindner (1828–1887) besondere Aufmerksamkeit. Nach Absolvierung der Universität wurde er Lehrer an einem Gymnasium und wurde wegen Verdachts des Atheismus von Böhmen in das damals südsteirische Celje (Cilli) strafversetzt, von wo er erst 1871 nach Böhmen zurückkehren konnte. Er bekam den Posten des Direktors am Lehrerinstitut in Kuttenberg (Kutná Hora), und 1882 – nach der Teilung der Prager Universität in eine tschechische und deutsche – wurde er schließlich zum Professor für Philosophie und Pädagogik ernannt. Während seines Aufenthalts in Celje schrieb er im Herbartschen Geiste drei Lehrbücher über die philosophische Propädeutik (*Lehrbuch der empirischen Psychologie nach genetischer Methode* (1858); *Lehrbuch der formalen Logik* (1861) und eine *Einleitung in das Studium der Philosophie* (1866)), die zahlreiche Auflagen erfuhren und in mehrere Sprachen übersetzt wurden. Lindner wurde dadurch zu einem der bekanntesten Herbartianer in Europa. Lindner interessierte sich vornehmlich theoretisch, aber auch praktisch für die Pädagogik.[22] Zur bis damals reichlich gepflegten und von den Herbartschen Grundlagen ausgehenden Völkerpsychologie leistete er durch die Arbeit: *Ideen zur Psychologie der Gesellschaft als Grundlage der Sozialwissen-*

22 Das bezeugen u.a. seine *Allgemeine Erziehungslehre* (1877) und die *Allgemeine Unterrichtslehre* (1877), beide parallel auf Deutsch und auf Tschechisch erschienen, sowie viele andere Schriften.

schaft (1871), einen bedeutenden Beitrag; diese Arbeit gilt zugleich als wichtigstes Werk der Soziologie in Herbartscher Tradition. Als Lindners Verdienst ist hier insbesondere zu sehen, dass ein dynamisches Element in die Herbartschen Konzeptionen eingebracht wurde, welches sich auf die Evolutionstheorie unter Bezug auf Darwin und Spencer stützt.

Von den jüngeren tschechischen Herbartianern trat zuerst Josef Dastich (1835–1870) hervor, dessen akademische Karriere durch eine tödliche Krankheit leider vorzeitig beendet wurde. Ab seiner Habilitation im Jahre 1861 begann man an der Universität mit systematischen Philosophievorlesungen in tschechischer Sprache, und Dastich hat neben Durdík große Verdienste an der Schaffung der tschechischen philosophischen Terminologie. Er verfasste auch die erste tschechische *Filosofická propedeutika* [*Philosophische Propädeutik*] (I–II, 1867). In dem der Psychologie gewidmeten Teil flocht er seine Erfahrungen ein, die er bei seinem Aufenthalt in Helmholz' experimentellen Labor in Heidelberg erworben hatte. In seinem Buch über die praktische Philosophie gab er zum ersten Mal in tschechischer Sprache eine systematische Interpretation der Herbartschen Ästhetik.

Eine glücklichere Karriere hatte sein etwas jüngerer Zeitgenosse Josef Durdík (1837–1902), der zur anerkannten Autorität des tschechischen Kulturlebens wurde und mit dessen Namen am öftesten der Begriff des „tschechischen Herbartianismus" verbunden wird. Durdík schrieb – ähnlich wie sein Lehrer R. Zimmermann – eine umfangreiche Herbartsche Ästhetik, *Všeobecná aesthetika* [*Allgemeine Ästhetik*] (1875), in der gewisse Forscher[23] die Anfänge der modernen semantischen Auffassung der Kunst zu sehen glauben. In seinem umfangreichen Werk finden wir Schriften, die alle von den Herbartianern gepflegten Hauptdisziplinen betreffen, neue Konzeptionen enthalten sie allerdings nicht, wenn man ihnen auch eine Reihe von originellen Bemerkungen zu Teilfragen nicht absprechen kann. Aus seiner Feder stammt auch die erste in tschechischer Sprache geschriebene zweibändige *Geschichte der neuzeitlichen Philosophie*, die man auch heute noch mit Interesse und Gewinn lesen kann.

Der Herbartianismus war die einflussreichste Richtung der tschechischen Philosophie des 19. Jahrhunderts. Sein langes Bestehen kann auch damit erklärt werden, dass er Anregungen aus anderen europäischen Geistesströmungen dieser Zeit aufnahm und positiv auf die Entdeckungen der Naturwissenschaften reagierte. In diesem Sinne ersetzte er die Rolle des Positivismus. Deshalb ist es nicht verwunderlich, dass der Comte'sche Positivismus erst gegen Ende des Jahr-

---

23 Vgl. Sus 1960, p. 776ff.

hunderts in Böhmen intensiver zur Geltung kam, obwohl die Herbartianer ihn kannten.[24]

2.3 Otakar Hostinský

Otakar Hostinský (1847–1910) ist der bedeutendste tschechische Ästhetiker des 19. Jahrhunderts. In idealem Maße vereinigte er in sich den Theoretiker und Praktiker. Er hatte ein außergewöhnliches Talent für das theoretische Denken, und von Jugend an – noch als Hörer von Dastich und Volkmann – bis ins reife Alter durchdachte er Fragen der allgemeinen Ästhetik. Parallel dazu schuf er eine Theorie der einzelnen Kunstarten, insbesondere der Musik, der bildenden und dramatischen Kunst, aber auch der Lyrik. Als Programmästhetiker proklamierte er die Grundsätze des künstlerischen Realismus. Als Kunstkritiker unterzog er die bedeutenden Kunstwerke seiner Zeit, wie z. B. die Musik von Richard Wagner oder Bedřich Smetana einer Wertung und verteidigte sie – wenn nötig.

Aus der Sicht der allgemeinen Ästhetik sind seine beiden Arbeiten von grundlegender Bedeutung. Die erste, *Über die Bedeutung der praktischen Ideen Herbarts für die allgemeine Ästhetik*, erschien 1883 in Prag (in tschechischer Sprache zwei Jahre früher). Hostinský lehnt hier Zimmermanns (und auch Durdíks) fünf höchste ästhetischen Formen ab (das Große, das Charakteristische, der Einklang, die Korrektheit, die Ausgleichung), die analog zu Herbarts fünf Grundsätzen des moralischen Verhaltens (die innere Freiheit, die Vollkommenheit, das Wohlwollen, das Recht, die Billigkeit) aufgestellt wurden. Er argumentierte, dass bei Herbart das Verhältnis der Ethik und der Ästhetik gerade umgekehrt sei, dass die sich auf den Willensverhältnissen gründende Ethik nur ein Sonderfall der Beziehung der ästhetischen Elemente sei, die konkret und in der Zahl unbegrenzt sind, und dass es somit keinen Grund gebe, sie auf eine gewisse Anzahl zu reduzieren und aus ihnen Abstrakta zu machen. Hostinský erblickte darin zu Recht die Gefahr des Dogmatismus und der Normativisierung der Ästhetik und somit ein Hindernis für den wissenschaftlichen und künstlerischen Fortschritt.

Der eingehenden Begründung dieses neuen Standpunkts widmete Hostinský das Buch: *Herbarts Ästhetik in ihren grundlegenden Teilen quellenmäßig dargestellt und erläutert* (1891), wo er alle Herbartschen Äußerungen zur Ästhetik sammelte, ordnete und mit einem eigenen Kommentar versah, um die Authentizität seiner Auffassung von der allgemeinen Ästhetik zu belegen und seine Interpre-

24 Eine umfangreiche Bibliographie und ausführliche biographische Beschreibungen der tschechischen Herbartianer findet sich im oben zitierten Buch Tretera 1989.

tation des Herbartianismus im Kampf um den künstlerischen Fortschritt zu verteidigen. Das Ergebnis dieser Verteidigung ist völlig überzeugend, und die abstrakt formalistische Interpretation Herbarts musste vor Hostinskýs und Herbarts „konkretem Formalismus" zurückweichen. Dies ist neben der Analyse der Bedingungen des Schönen, die in anderen Studien durchgeführt wurden, der größte Beitrag Hostinskýs zur allgemeinen Ästhetik. Der Umstand, dass es sich hier eigentlich um die Erneuerung des ursprünglichen Sinnes der Herbartschen Konzeption handelte, konnte diesem Verdienst keinen Abbruch tun. Die Hauptaussagen Herbarts benötigten nämlich keine umfangreichen Kommentare, um deren theoretischen Wert ersichtlich zu machen.

In einem Punkt aber ging Hostinský mit Herbart nicht konform. Er warf ihm vor, ungerechtfertigterweise die Mimesis oder – wie die Herbartianer zu sagen pflegten – „das Charakteristische" aus der allgemeinen Ästhetik auszuschließen, und stimmte Zimmermann zu, der sie für eine ästhetische Kategorie hielt. Es handelte sich hier um nichts Geringeres als die Beziehung zwischen Schönheit und Wahrheit, um die Wahrhaftigkeit der Abbildung. Hostinský konnte es als überzeugter Anhänger des damaligen Realismus nicht ertragen, dass ein solch bedeutendes Verhältnis in die bloße Kunsttheorie fallen sollte, wie es Herbart lehrte (obwohl Herbart die künstlerische Bedeutung nicht unterschätzte). Wenn auch Hostinský seinen Standpunkt mit scharfsinnigen Argumenten[25] untermauerte, obwohl er dieses Verhältnis als sehr lose verstand und jedwede mechanische Nachahmung ablehnte, scheint es, dass die Entwicklung der modernen Kunst eher Herbart Recht gibt.

Hostinský ergänzte sein System der allgemeinen Ästhetik während der ganzen Dauer seiner aktiven wissenschaftlichen Tätigkeit immer wieder, was die Aufzeichnungen von seinen Vorlesungen an der Universität bezeugen. Obgleich er ihr nie definitive Textgestalt gab, können wir Hostinský mit Recht als den bedeutendsten Systematisierer, als kreative Persönlichkeit der Herbartschen Ästhetik und zugleich als deren Vollender betrachten.[26]

3 Herbartsche Initiativen. Schlussfolgerungen

Die gedanklichen Zusammenhänge von Herbarts Anschauungen mit der weiteren Entwicklung der Philosophie und der Wissenschaftsfächer, die damals im Rahmen der Philosophie betrieben wurden und heute eigenständige Wissen-

25 Siehe Nejedlý 1921, p. 400–402.
26 Siehe dazu u. a. Zumr 1958, p. 301ff.; Zumr 1982, p. 77ff.; Volpicelli 1992, p. 625ff.

schaften darstellen, sind bei weitem nicht in allen Einzelheiten erforscht. Immer wieder werden neue Motive entdeckt, deren Filiation deutlich auf Herbart hinweist. Oft kann man natürlich die tatsächliche Genealogie des jeweiligen Gedankens nicht konkret belegen, insbesondere wenn ihr neuer Entdecker sich nicht explizit zum Urheber bekennt. Dies muss nicht immer durch das Bestreben, die eigene Originalität nicht herabsetzen zu wollen, motiviert sein. Manchmal handelt es sich nur um eine unzureichende Kenntnis der Geschichte des Denkens, manchmal vielleicht auch um Scheu, sich zu einem Autor zu bekennen, der durch die spätere Entwicklung der allgemeinen Meinung oder durch die bewussten Bemühungen der Nachfolger um eine radikale Distanzierung diskreditiert wurde. Im Falle Herbarts tritt hier noch ein weiterer Umstand hinzu: Nach mehr als einem halben Jahrhundert von tausend Wissenschaftsadepten, insbesondere in Mitteleuropa, drang er in die Gedankenwelt mittels philosophischer Mittelschulpropädeutiken ein, die in Herbarts Geiste verfasst wurden. Viele von diesen Gedanken wurden zu *loci comunes*, gingen in das allgemeine Bewusstsein über und verloren den Zusammenhang mit ihrem Urheber. Und so stehen die Philosophiehistoriker vor der Aufgabe, die versteckten gedanklichen Zusammenhänge ausfindig zu machen und das kollektive menschliche Gedächtnis zu erneuern.

Die Ideenbeziehungen haben in der Regel in den Fällen klarere Konturen, in denen sie die Autoren gegenseitig polemisch definieren. Was Herbart betrifft, so waren dies Polemiken, die er mit seinen Zeitgenossen führte oder die diese mit ihm führten. Darin kommen Namen vor wie Fichte, Schelling, Hegel, Schopenhauer, Fries, Beneke, Bolzano, Trendelenburg und andere, weniger bekannte. Aber auch hier lässt sich manchmal eine größere Verwandtschaft finden, als der polemische Enthusiasmus zulassen würde.

Einfacher ist die Situation bei den Wissenschaftlern, die die Inspiration durch Herbartsche Gedanken selbst konstatiert haben, wie die bereits zitierten Carl Friedrich Gauß, Wilhelm Wundt, weiters Georg Friedrich Bernhard Riemann, der Herbart für den Vorgänger seiner Raumtheorie hielt, oder Moritz Lazarus, einer der Begründer der Völkerpsychologie, der Herbart als Newton der Psychologie bezeichnete.[27] In diese Kategorie gehört auch die Beziehung des wichtigsten Repräsentanten der Prager ästhetischen strukturalistischen Schule, Jan Mukařovský, zur Herbartschen Ästhetik, der sich einer gewissen genetischen Verwandtschaft der beiden Schulen bewusst war (mittels der Tradition der tschechischen Ästhetiker Otakar Hostinský und Otakar Zich); die allseitigen und tieferen Zusammenhänge wurden durch die philosophiegeschichtliche For-

27 Siehe Pettoello 1988, p. 140–142.

schung bewiesen.[28] Den Bereich der Pädagogik lassen wir beiseite, da sie bis jetzt am besten erforscht wurde und in der Pädagogik die ursprünglichen Anschauungen Herbarts maximal aktualisiert wurden.

Im Bereich der Arbeitshypothesen, die sich zwar manchmal auf authentische Aussprüche stützen, aber kritisch untersucht werden müssen, bleibt ein Verhältnis zu Herbart bei Denkern wie Franz Brentano (negative Beziehung zum Hegelianismus), Edmund Husserl (Logik auf antipsychologischer Grundlage), Ernst Mach (die Empfindungselemente), Christian von Ehrenfels (der Gestaltbegriff),[29] bei den Philosophen des Wiener Kreises (Philosophie als Bearbeitung der Begriffe), aber auch bei Künstlern wie Adalbert Stifter[30] usw. Und nicht zuletzt stützen eindeutige Belege Studien, die einen durch Lindner vermittelten Einfluss der Herbartschen Psychologie auf das Entstehen von Freuds Psychoanalyse nahelegen.[31] Es würde aber den Rahmen dieser Arbeit sprengen, auf derartige Zusammenhänge näher einzugehen.

Aus unserer kurzgefassten Übersicht über die Entwicklung des Herbartianismus in den böhmischen Ländern geht hervor, dass es hier schon im 19. Jahrhundert gelang, auf internationaler Basis und in Zusammenarbeit zwischen tschechischen und deutsch-österreichischen Persönlichkeiten, in Mitteleuropa eine fruchtbare intellektuelle Atmosphäre zu schaffen, die in mannigfacher Weise bemerkenswerte Beiträge zu und Anregungen für die weitere Entwicklung von Wissenschaft, Kunst und Bildung leistete.

Literatur

Dastich, Josef (1867), *Filosofická propedeutika* [*Philosophische Propädeutik*], 2 Bde., Prag.
Durdík, Josef (1875), *Všeobecná aesthetika* [*Allgemeine Ästhetik*], Prag.
Hemecker, Wilhelm (1987), „Sigmund Freud und die Herbartianische Psychologie des 19. Jahrhunderts", in: *Conceptus*, XXI, Nr. 53/54, p. 217–231.
Hostinský, Otakar (1891), *Herbarts Ästhetik in ihren grundlegenden Teilen quellenmäßig dargestellt und erläutert*, Hamburg.
Jánský, Karel (1958), *K. H. Mácha ve vzpomínkách současníků*, Prag.
Lentze, Hans (1962), *Die Universitätsreform des Ministers Graf Leo Thun-Hohenstein*, Wien.
Lindner, Gustav Adolf (1858), *Lehrbuch der empirischen Psychologie nach genetischer Methode*, Cilli.
Lindner, Gustav Adolf (1861), *Lehrbuch der formalen Logik*, Graz.

28 Siehe Anmerkung Nr. 26.
29 Zu Ehrenfels siehe die Studie von Pauza 1995, p. 935.
30 Vgl. z. B. Mühlher 1948, p. 217ff.
31 Vgl. Hemecker 1987, p. 217ff.

Lindner, Gustav Adolf (1866), *Einleitung in das Studium der Philosophie*, Wien.
Lindner, Gustav Adolf (1871), *Ideen zur Psychologie der Gesellschaft als Grundlage der Sozialwissenschaft*, Wien.
Lindner, Gustav Adolf (1877), *Allgemeine Erziehungslehre*, Wien.
Lindner, Gustav Adolf (1877), *Allgemeine Unterrichtslehre*, Wien.
Mühlher, Robert (1948), „Ontologie und Monadologie in der österreichischen Literatur des XIX. Jahrhunderts", in: *Die österreichische Nationalbibliothek*, hrsg. v. J. Stummvoll, Wien, p. 217ff.
Nejedlý, Zdeněk (1921), *Otakara Hostinského esthetika*, Prag.
Pauza, Miroslav (1995), „Koncepce ‚tvarových kvalit' Christiana von Ehrenfelse", *Filosofický časopis*, Bd. 43, No. 6, Prag, p. 935–958.
Pettoello, Renato (1988), *Introduzione a Herbart*, Rom/Bari.
Smetana, Augustin (1848), *Die Bedeutung des gegenwärtigen Zeitalters*, Prag.
Smetana, Augustin (1850), *Die Katastrophe und der Ausgang der Geschichte der Philosophie*, Hamburg.
Smetana, Augustin (1865), *Der Geist, sein Entstehen und Vergehen*, Prag.
Steiner, Rudolf (1925), *Mein Lebensgang*, Dornach.
Sus, Oleg (1960), „Sémantické problémy umění u Josefa Durdíka", in: *Filosofický časopis*, Bd. 8, Prag 1960, p. 776ff.
Tretera, Ivo (1989), *J. F. Herbart a jeho stoupenci na pražské univerzitě*, Prag.
Volpicelli, Ignazio (1992), „‚Formalismo' e molteplicit, estetica in J. F. Herbart", in: *J. F. Herbart. 1841–1991*, ed. Renato Pettoello (= *Problemi della Pedagogia*, XXXVIII, N. 6), p. 625ff.
Winter, Eduard (ed.) (1975), *Robert Zimmermanns Philosophische Propädeutik und die Vorlagen aus der Wissenschaftslehre Bernard Bolzanos. Eine Dokumentation zu Geschichte des Denkens und Erziehung in der Donaumonarchie*, in: *Österreichische Akademie der Wissenschaften*, Sitzungsberichte, philosophisch-historische Klasse, Bd. 299, Abh. 5, Wien.
Zimmermann, Robert (1847), *Leibniz' Monadologie mit einer Abhandlung über Leibnizens und Herbarts Theorien des Wirklichen Geschehens*, Wien.
Zimmermann, Robert (1848), „Entwurf der Grundzüge des öffentlichen Unterrichtswesens in Österreich", in: *Wiener Zeitung*, Nr. 197–200, 18. bis 21. Juli.
Zimmermann, Robert (1849), *Über den wissenschaftlichen Charakter und die philosophische Bedeutung Bernard Bolzanos*, Wien.
Zimmermann, Robert (1849), *Leibniz und Herbart, eine Vergleichung ihrer Monadologien*, Wien.
Zimmermann, Robert (1852a), *Philosophische Propaedeutik für Obergymnasien*, Prag; zweite Aufl. Wien, 1860; dritte Aufl. Wien, 1867.
Zimmermann, Robert (1852b), *Was erwarten wir von der Philosophie? Ein Vortrag beim Antritt des ordentlichen Lehramts der Philosophie an der Prager Hochschule*, gehalten am 26. April 1852, Prag.
Zimmermann, Robert (1852c), *Das Rechtsprinzip bei Leibniz*, Wien.
Zimmermann, Robert (1856), *Über das Tragische und die Tragödie*, Wien.
Zimmermann, Robert (1859), *Schiller als Denker*, Prag.
Zimmermann, Robert (1888), „Philosophie und Philosophen in Österreich", in: *Österreich-Ungarische Revue*, Neue Folge, 6. Band, Oktober 1888 bis März 1889.
Zumr, Josef (1958), „Theoretické základy Hostinského estetiky", in: *Filosofický časopis*, 6, Prag, p. 301ff.

Zumr, Josef (1982), „Hostinský a Herbart“, in: *Pocta Otakaru Hostinskému*, ed. R. Pečman, Brünn, p. 77ff.

Carole Maigné

Formwissenschaft and Phantasy: Robert Zimmermann's Formal Aesthetics

Abstract: This paper examines the status of the image in Herbartian philosophy and its legacy in Zimmermann's Formwissenschaft. To work on the question of the image imposes to articulate philosophy and psychology, to articulate anti-psychologism in aesthetics and the production of aesthetic effects understood according to the threshold of consciousness that Herbart installs. Zimmermann's so called "logicism" must therefore always also take into account the dynamics of representation, because Formalism is not conceived in rupture from scientific psychology. Our hypothesis is that Zimmermann's Formwissenschaft offers an original theory of imagination that extends Herbart's Cultursystem: the image in his text is not only a problem of visibility but also of shaping the cultural world. The theory is embodied in the concept of Phantasy.

Introduction

Our contribution will focus on the concept of "general aesthetics" (*Allgemeine Aesthetik*) that Zimmermann stands for, namely the *Formwissenschaft*. The science of form aims at an objectivity that breaks with all metaphysics of art but also with all subjectivism: as such, art is neither a dialectical moment in the history of the Spirit, nor the expression of a subjective intuition inaccessible to the concept. Zimmermann is indeed a representative of the formalism of visibility, since what counts are relations between artistic elements, "form" being in Herbartism a question of relations (*Verhältnisse*). What matters is the way in which the artistic elements (pictorial, musical, etc.) fit together. It is also a formalism of judgment, because general aesthetics is the science of what pleases or displeases, in accord ance with Herbart's definition of aesthetics in § 8 of his *Lehrbuch zur Philosophie*:

Note: I warmly thank Katherine Arens and Denis Fisette for their attentive proofreading.

Carole Maigné, University of Lausanne
carole.maigne@unil.ch

https://doi.org/10.1515/9783110747324-007

Aesthetics is the science of those concepts that bring about an addition in our conceiving, consisting in judging whether we approve or disapprove.[1] Two features characterize the aesthetic judgment: it is partly logical, without being totally logical (it does not respond to a *logische Verdeutlichung*), it responds to another legality; it is not empirical, since liking or disliking does not depend on the empirical reality of the object.

Zimmermann continues Herbart's words and seeks the full autonomy of aesthetic science. The objectivity of Formwissenschaft is not that of knowledge, aesthetic science is not a theoretical science stricto sensu.[2]

> Since the aesthetic concepts are mere images and not copies, it follows automatically that the addition cannot be determined by the theoretical correctness and validity, that is to say, by the truth and falsity, of their conceptual content. The latter depends on the relationship of the representation to the *thing*, whereas the aesthetic addition depends on the mere *representation*. "Knowledge", "imagination", history and fairy tale are therefore equally valid in relation to the aesthetic addition. [...] The theoretical and the aesthetic view of the world have different interests.[3]

The aesthetic judgment does not depend on the empirical existence of the object, nor on its truth or falsity. Aesthetics as a science thus claims the autonomy of the beautiful as to the existence of what is or what can be, and aesthetics does refuses mimesis. It is a problem of form and adherence to form. Aesthetics as a science of "fundamental aesthetic forms" (*Ästhetische Grundformen*)[4] thus works with

1 Herbart 1837, § 8, p. 52. Full quote is: "Noch gibt es eine Klasse von Begriffen, die mit den vorerwähnten darin übereinkommen, daß bei ihnen das Denken nicht bei bloßer logischer Verdeutlichung still stehen kann; die sich aber dadurch unterscheiden, daß sie nicht, gleich jenen, eine Veränderung notwendig machen, wohl aber einen Zusatz in unserem Vorstellen herbeiführen, der im Urteile des Beifalls oder Mißfallens besteht. Die Wissenschaft von solchen Begriffen ist die Ästhetik. Mit der Kenntnis des Gegebenen hängt sie ihrem Ursprunge nach nicht weiter zusammen, als insofern wir dadurch veranlaßt waren, uns Begriffe vorzustellen, welche, ohne alle Rücksicht auf ihre Realität, den Beifall oder das Mißfallen erwecken."

2 Zimmermann 1865, § 22.

3 Zimmermann 1865, § 25: "Da die ästhetischen Begriffe bloße Bilder und nicht Abbilder sind, so folgt von selbst, dass der Zusatz nicht durch die theoretische Richtigkeit und Gültigkeit, d. h. nicht durch Wahrheit und Falschheit ihres Begriffsinhalts bestimmt sein könne. Letztere hängt vom Verhältnis der Vorstellung zur *Sache*, der ästhetische Zusatz dagegen von der bloßen *Vorstellung* ab. 'Erkenntnis', 'Einbildung', Geschichte und Märchen gilt daher in Bezug auf den ästhetischen Zusatz gleich. [...] Die theoretische und die ästhetische Weltansicht haben verschiedene Interessen."

4 Zimmermann 1865, § 72.

"normal images" (*Normalbildern*),[5] of which it proposes a typology. These normal images are stabilized images in five standard judgments.[6]

But how do these fundamental aesthetic forms manifest themselves? How is their universal and necessary character imposed? Zimmermann replies that the image and the addition of likes or dislikes coincide in a "perfect conceiving" (*Vollendetes Vorstellen*),[7] and it is this coincidence that makes it imposing:

> Because this is nothing else than the necessary and inevitable effect of the perfect representation of a composite picture, by virtue of which the subject does not judge but is *made* to judge by the object, so the predicate, namely the feeling of pleasure and displeasure, is in this case nothing other than the expression of the tension between the parts of the composite image, i. e. the form itself, therefore subject and predicate are *identical*, and consequently the judgment is *evident*.[8]

Zimmermann presupposes an experience of judgment whose accomplished character lies in being totally obvious because it is desubjectified and necessary because it is constrained by the way the object is arranged (not by its nature as an object). If "the aesthetic *quid* of what pleases or displeases is thus basically a *how*",[9] formalism is a theory of the modalities of what pleases or displeases in arrangements composed of aesthetic elements, captured in images and as images. There is thus an immanence of form in the aesthetic judgment that guarantees its objectivity.

The critique of logicism often raised against Zimmermann is rooted here, to the point that Zimmermann does defend the paradox of an a-aesthetic aesthetics.[10] For how can we explain that these images do stabilize and impose themselves? Are likes and dislikes a-temporal, a-historical and so cut off from any individual emotion that one is "durch das Objekt urteilen gemacht"? Must we

5 Zimmermann 1865, p. VIII.

6 Zimmermann calls them characteristic, agreement, correction, comparison and compensation, our purpose here is not to discuss them.

7 Zimmermann 1865, § 45 and 51.

8 Zimmermann 1865, § 59: "Da dieses nichts Anderes ist als der notwendige und unvermeidliche Effekt der vollendeten Vorstellung eines zusammengesetzten Bildes, vermöge dessen das Subjekt nicht urteilt, sondern durch das Objekt urteilen *gemacht* wird, so ist das Prädikat, nämlich das Lust- und Unlustgefühl in diesem Fall nichts Anderes als der Ausdruck der Spannung zwischen den Teilen des zusammengesetzten Bildes, d. h. die Form selbst, sonach Subjekt und Prädikat *identisch* und folglich das Urteil *evident*."

9 Zimmermann 1865, § 58: "das *Was* des ästhetischen Gefallenden und Missfallenden demnach im Grunde ein *Wie* sei".

10 I refer here to the analysis of Wiesing 2001 and Henckmann 1993.

therefore admit that these forms are without reason ("grundlos")[11] as much as without object ("gegenstandslos")? Our contribution would like to show that Zimmermann is not satisfied with a logicism of images that appeal without reason and without object. Formal and abstract evidence is modulated by a concept that seems neglected: that of *phantasy*, which is neither subjective imagination (*Einbildung*), nor a return to the psychology of faculties (according to Herbart's critique). In our opinion, he does insist on the non-objectuality of images to go beyond Kant's transcendental definition of imagination. The Herbartian formalism is constructed thanks to a psychology of the depths, that of the threshold of consciousness, and not against it: it is because psychology has become science that formalism can be erected. The anti-psychologism of aesthetic judgment does not deny the psychology of the threshold, and this threshold is individual as well as collective. The matter of aesthetic judgments is indeed vivid representations-power (*lebendige Vorstellungskräfte*),[12] and the image is itself a question of threshold in Herbartism: its obviousness is secondary, it results from a dynamic process of representations, therefore from an active sedimentation. Zimmermann's latest work, his *Anthroposophy*, shows that adherence to objectivity is a matter of attention and education: this opens the concept of aesthetics to a theory of culture, understood as the set of its objective images that please or displease and to which one adheres according to a collective objectivity whose rules are immanent. The image in this formal aesthetic goes beyond the paradigm of seeing: everything becomes an image in *Anthroposophy*, a park, fashion, a channel, because everything is crystallized into an active cultural form.

1 The Image Threshold and the Aesthetic Judgment According to Herbart

To understand the status of image and phantasy in Zimmermann, it seems important to return to Herbart. Herbart proposed an original approach to the image through his eidolology (*Eidolologie*), namely his theory of the subject. The term is not in itself innocent: it is precisely a questioning of the theory of *eidola* coupled with a critique of the subject. The image is not unified by resemblance

11 Zimmermann 1865, § 77: "Sie sind *Grundformen*, welche gefallen oder missfallen, eben deshalb auch *grundlos* gefallende und missfallende Formen. Ihr Gefallen und Missfallen, weil es nur das Werk der im Zusammen befindlichen Vorstellungen selbst ist, heisst *unbedingt*."
12 Zimmermann 1865, § 69.

to an object nor is it unified by a subject which would be fixed once and for all. An image is a series of representations, due to their dynamic; it involves the psychology of the threshold of consciousness that dissociates representation and consciousness of this representation. Herbart therefore introduces latency, opacity, mobility and fluidity into his definition. The image is no longer thought of so much from the model of sight as from the dynamics of representations, the eye basically losing the pre-eminence in the definition of the visible of and in the image. This induces an astonishing regime of visibility: a visibility where the paradigm of seeing is subjected to a process of "making visible", where the evidence of the image is conquered and not acquired. Aesthetic obviousness is thus possible against the background of a psychological process that the subject does not fully master.

Knowledge by Herbart is not a copy (*Abbild*) of things, so consciousness is not a gallery of images (*Bildersaal*) where "all kinds of paintings of the world and their changing shapes are together"[13]. Things are not transported into us in images similar to these things: this is the old *eidola* theory that is outdated. Because sensation is not the reflection of something outside of us but of the self-conservation of the soul, ontology has become an analytic of understanding since Kant. It is a common error to think that things give themselves spontaneously to us, in an obviousness that is traditionally granted to the visible: "Would we really let the old democritic Eidola flutter in the air?",[14] believing that these *eidola* bear the resemblances to the things they claim to present to us? Herbart finds here one of Sartre's primary demands in *L'imaginaire*: the image is not something in our mind. But where Sartre thought in terms of the aim of consciousness, of intentionality, Herbart chooses another path: the disengagement between representation and consciousness of that representation, because the I does not accompany all my representations. The differentiation of the representations does not come from a view of the eye or of the soul thought to be like an eye, but from the relations that the representations create with each other:

> All representations in the narrower sense, that is, those that are a picture of some object, whether real or apparent or fictional, are webs of series that are passed through in a rapid succession imperceptibly continuing. The swing through the partial representations leaves

13 Herbart 1829, SW VIII, § 323, p. 229: "allerley Gemälde von der Welt und ihren wechselnden Gestalten beysammen sind".
14 Herbart 1829, SW VIII, § 309, p. 207: "Würden wir wirklich die alten Demokritischen *Eidola* in der Luft herumflattern lassen?"

> behind an overall impression, which at any moment can be set in motion again by the slightest inducement.[15]

If the picture (*Bild*) is therefore neither *Urbild* nor *Abbild*, it is a process. All of Herbart's pictorial analogies insist on the dissociation between representation and object of representation: this is the threshold of consciousness that his psychology sets up. Consciousness of oneself and of one's objects corresponds to a certain "Quantum of self-observation" (*Quantum der Selbstbeobachtung*) that develops to a certain degree and under certain circumstances. As Herbart's readers know, attention is not self-evident, it fluctuates, it can even disappear, it is of degrees.[16] However, it is not purely passive because it also responds to a logic of expectation:

> Stimulated expectation promotes apperception; thus we observe a play in which the very beginning of it sets in motion a number of representations as to how the play *might* go on; with them the *real* course of events then enters into all kinds of relations of inhibition and fusion – the same now also happens *internally*; without the conceptions being given *from outside*.[17]

We also anticipate what is seen: representations arise, without necessarily stabilizing in images, some others inhibit them, some are knotted, others are not. Herbart concludes: "Here the inner sense is present, even if the apperceptive representation is not always assigned to us as *our* representation"[18]. The image therefore occurs both within me and outside me, with me and without me, belonging to me (since no one else produces it for me, in my place) and not belonging to me (these movements of representations have their own logic which is not that of the consciousness I have of them).

---

15 Herbart 1825, SW V, § 100, p. 417: "Alle Vorstellungen im engern Sinne, das heisst, solche, die ein Bild sind von irgend einem, gleichviel ob wirklichen, oder scheinbaren, oder erdichteten Gegenstande, sind Gewebe von Reihen, die in einer schnellen Succession unmerklich fortfliessend, durchlaufen werden. Der Schwung durch die Partial-Vorstellungen lässt einen Gesammt-Eindruck zurück, der jeden Augenblick auf die geringste Veranlassung wieder in irgend eine innere Bewegung gerathen kann."

16 Herbart 1822a, SW V.

17 Herbart 1825, SW VI, § 125, p. 142f.: "Angeregte Erwartung befördert die Apperception; so beobachten wir ein Schauspiel, indem gleich der Anfang desselben eine Menge von Vorstellungen in Bewegung bringt, wie das Stück wohl fortgehen *könnte*; mit welchen alsdann der *wirkliche* Verlauf in allerley Verhältnisse der Hemmung und Verschmelzung eintritt. – Dasselbe nun geschieht auch *innerlich*; ohne dass die Auffassungen *von außen* gegeben werden."

18 Herbart 1825, SW VI, § 125, p. 143: "Hier ist der innere Sinn vorhanden, wenn auch die appercipirte Vorstellung nicht immer als *unsere* Vorstellung Uns zugeeignet wird."

"*Who* is then the observer, *who* is being observed?",[19] a question which is reinforced in § 126:

> [...] but in the case of inner perception, where both, the apperceived and the apperceptive, are inner, one can well stand and ask: which representation is *being appropriated* here and which is the *appropriating* one?[20]

An image is at the same time deeply impersonal and personal: an image is impersonal because there are representations in me that I didn't want, nor chose, nor decided on, and is personal because it is a sedimentation of my own. What is also striking is the reversibility between active and passive, between determinant and determined, between subject and object. "Having" an image is sometimes an image "for", seeing and being-seen seem to go hand in hand. In any case, it is always having an image by degrees, observation is a matter of perception and attention: we must therefore be available to the image, even if we are the producers. The evanescence of the image and its crystallization are the two poles of a psychologization of the image according to mechanisms of inhibition (*Hemmung*) and repression (*Verdrängung*) that never cease to act.

The image is learned, as one learns to perceive, to feel, by following an *ABC der Anschauung*.[21] Eidolology therefore assumes that "we live but in relations and need nothing more"[22]. The objectivity of our represented world is always mediated and acquired: "There is no direct knowledge here, only indirect knowledge. This means in other words: the existing does not get represented by itself in the soul."[23] There is indeed an objective appearance valid for all, but eidolology also tells us that, from a logical point of view, the general cannot benefit from an absolute position: according to the eidolology there are therefore only abbreviations of singular things, abbreviation and condensation serving usage, it is therefore advisable not to reify the general, not to give it a meaning or a status which it does not have ("the general is only abbreviation for convenience without any

19 Herbart 1825, SW VI, § 125, p. 140: "*Wer* ist alsdann der Beobachtende, und *wer* wird beobachtet?"

20 Herbart 1825, SW VI, § 126, p. 143: "Aber bei der innern Wahrnehmung, wo beides, das Appercipirte und das Appercipirende, innerlich ist, kann man wohl anstehen und fragen: welche Vorstellung *wird* hier *zugeeignet*, und welche ist die *zueignende*?"

21 Herbart 1802; Maigné 2018, p. 75–88.

22 Herbart 1829, SW VIII, § 328, p. 239: "wir leben einmal in Relationen, und bedürfen nichts weiter".

23 Herbart 1829, SW VIII, § 329, p. 241: "Hier giebt es kein unmittelbares Wissen, sondern nur ein mittelbares. Das heisst mit andern Worten: das Seyende bildet sich in der Seele nicht von selbst ab."

meaning of its own")[24]. The abbreviation of course refers to Leibniz, to the blind and mechanical thinking within us. But since Herbart does not have a metaphysics of pre-established harmony, there is no *a priori* adjustment of meaning, there is only experience and its laws. A darkness is at the very heart of the clarity of consciousness, something that does not reach the conceptual, *de facto* and *de jure*. The abbreviation means that something does not reach the clear and obvious formulation of the concept. The metaphysics is "science of the comprehensibility of experience" (*Wissenschaft von der Begreiflichkeit der Erfahrung*),[25] it is not explanation of experience. And so, in a sense that is not, however, Fichtean, the image never quite translates itself into a concept: not because it exceeds the concept by the intuition of something that exceeds it, but because, on the contrary, the flow of representations does not rely on an all-powerful subject, does not presuppose either a philosophy of the absolute or an omnipotent intuition.

As Lambert Wiesing has shown,[26] Herbart deploys an original concept of visibility (*Sichtbarkeit*): he releases an immanent reality of relations on the very surface of the image, autonomous from the objective content (*gegenständlich*). Formal aesthetics is dedicated to the immanent relations that the parts of the image have with each other on the surface of the image, which is not simply a concern for the composition but for the reciprocal differentiation of the elements. Herbart participates in the quest for the "pure visibility" of the 19th century, for an image (*Bildlichkeit*) that should be distinguished from any other object. But another aspect of this quest for visibility must be emphasized: psychology, as the science of the threshold of consciousness, knows that the image is irreducible to the concept, irreducible to the clarity of the idea, irreducible to its logical formulation. The notion of threshold prevents logical transparency: metaphysical eidolology is inseparable from a scientific psychology, since psychology as a science is "applied metaphysics". And the latter proposes a rich phenomenology of the unintentionality of the image, a phenomenology of the appearance of appearance that rests on the threshold of consciousness. "The eye at rest sees no space":[27] the law of order proper to visibility is that of transition (*Übergang*),[28] that of the pro-

24 Herbart 1829, SW VIII, § 329, p. 240: "das Allgemeine ist nur Abbreviatur, zur Bequemlichkeit, ohne irgend eine eigene Bedeutung."

25 Herbart 1828, SW VII, § 81, p. 132.

26 Wiesing 2008.

27 Herbart 1825, SW VI, § 111, p. 91: "Das ruhende aber Auge sieht keinen Raum."

28 Herbart 1825, SW VI, § 111, p. 91. Lambert Wiesing insists in his analysis of Zimmermann on this term "transition", because he sees in it all the descriptive potential of Herbartian aesthetic formalism, particularly in Alois Riegl.

gressive circulation of the eye between points, because the idea of distance is elaborated not by an *a priori* knowledge of the line, but by the path of the eye. To the eye, one must add the hand, or more broadly the body that moves: the image does not emerge from the eye alone, but from everything that organizes space, and all the body's movements participate in it. As Stout noted as early as 1891, attention is a motor process that sustains apperception:

> This means that the novelty, being in conflict with our preformed ideas, obstructs apperception, and so occasions a motor activity by which the apperceived presentation is invested with pre-dominant intensity and distinctness.[29]

We find ourselves thus, he continues in § 6 of his article, faced with two concomitant options, cooperation and competition:

> In so far as the mind is a unity, it tends to become affected as a whole by changes taking place in any of its component elements. For this reason an appercipient system tends to excite other systems in a degree varying directly with the intimacy of its connexion with them. The apperceptive activity of one group communicates to others a wave of ex-citation by which they are prepared to become in their turn appercipient [...]. On the other hand, every ideal group in the exercise of its apperceptive function tends to debar all other groups from becoming appercipient, excepting such as are at the moment capable of combining with it in the same systematic activity.[30]

The taste is stability, maintenance, as a chord is maintained at the piano: "Its verdict is a continuous sound, which does not fade, until the image is pulled away."[31] The taste does not merge with the universality of the true, it is also a fragile image, because there is always a need to control. The stability of this image is that of the possible detachment from a completed representation: it is therefore an evidence, freed from the weight of feeling, because aesthetic judgment is not affective. Judgment must be stripped of what encumbers it. The evidence is precisely in this fullness that comes neither too early nor too late (once the attention has been blunted).

> In order to have a pure judgment of taste at all, pay attention to the changeable nature of the states into which it puts the mind. What is changeable is to be separated; it cannot be essential to taste. But the perception of the object must remain in its sharpness, so that judgment can be made. Neither the first nor the last sensations aroused by a work of art are

29 Stout 1891, § 5, p. 33.
30 Stout 1891, § 6, p. 33f.
31 Herbart 1808, SW II, p. 342: "Sein Spruch ist ein anhaltender Klang, der nicht verstummt, als bis etwa das Bild hinweggezogen wird."

> purely aesthetic; the former because the object is not yet fully grasped, because the mass is still pressing; the latter because the attention is tired and waning.[32]

The image in question here is therefore not a simple perceived image (*aesthesis*). But it is not that it is an-aesthetic, rather the judgment made about it is only aesthetic, and aesthetics is not psychological. Aesthetics tears us away from the perceived, from the felt.

> Aesthetic judgments can only be made about relations; and it is the necessary proof of their correctness that the value of relations disappears as soon as the terms are separated; but reappears again when they are recombined. The proof shows what was important; namely, that no feeling of pleasure or displeasure (and even less desire) has interfered; rather, the mere summarizing observation has recognized the value of the object in which the relation lies. The dissection of such an object can only be a logical abstraction; as if one wanted to look at the colored parts of a painting individually, whereas it is precisely the given combination and arrangement of these parts that constitutes the essence of the painting.[33]

2 Phantasy, a Key Concept of Zimmermann's *Formwissenschaft*

The image here is therefore an arrangement: the image holds the immanent laws of its visible development, only its surface counts. Zimmermann is here in full agreement with Herbart. The logic of the image is conceived as a logic of relations on its surface: Lambert Wiesing qualifies Zimmermann's formalism as logicism,

32 Herbart 1808, SW II, p. 340: "Um überhaupt ein Geschmackurteil rein zu haben, achte man auf das Veränderliche der Zustände, in welche es das Gemüt versetzt. Dies Veränderliche sondere man ab; es kann dem Geschmack nicht wesentlich sein. Aber die Auffassung des Gegenstandes muss bleiben in ihrer Schärfe, damit geurteilt werden könne. Weder die ersten, noch die letzten Empfindungen, welche ein Kunstwerk erregt, sind die rein ästhetischen; jene nicht, weil der Gegenstand noch nicht vollkommen gefaßt ist, weil die Masse noch drückt; diese nicht, weil die Aufmerksamkeit ermüdet ist und schwindet."

33 Herbart 1836, SW X, p. 318f.: "Ästhetische Urteile können nur über Verhältnisse ergehen; und es ist die notwendige Probe ihrer Richtigkeit, dass der Wert der Verhältnisse verschwindet, sobald man die Glieder vereinzelt; hingegen wieder hervortritt bei erneuerter Zusammenfassung. Die Probe zeigt das, worauf es ankam; nämlich, dass sich kein Gefühl des Angenehmen oder Unangenehmen (und noch weniger ein Begehren) eingemischt hat; vielmehr die blosse zusammenfassende Betrachtung den Werth des Gegenstandes erkannt hat, in welchem das Verhältnis liegt. Die Zergliederung eines solchen Gegenstandes kann übrigens nur eine logische Abstraction seyn; wie wenn man die farbigen Stellen eines Gemäldes einzeln betrachten wollte, während gerade die gegebene Verbindung und Anordnung dieser Stellen das Wesentliche des Gemäldes ausmacht."

which manifests its limit as well as its fecundity.[34] We defend the idea that what he proposes is more a theory of the imagination than of the image. His problem is not only the visibility in the modalities of its being seen, it is also the morphology of the very act of creating that these modalities indicate; visibility is one of the possibilities. There is a recurring term in Zimmermann's work that reinforces our argument: that of *phantasy*. This term appears several times in Book I of the *Allgemeine Aesthetik*, strategically, and this is all the more remarkable since it is not strictly Herbartian. There is a discrepancy here, a shift, a shift that points towards Eduard Hanslick, whose place in Zimmermann's career cannot be overemphasized. If the critical literature has often insisted on the fact that Hanslick is not strictly a Herbartian,[35] based on the fact that his formalism is not "strict", we propose to turn the argument in the other direction: Zimmermann is also inspired by Hanslick, and he too tries to think of a dynamism of form, which is proved by the term morphology, which he takes up on his own.

It is difficult to see this as a coincidence when one considers the recurrent way in which the two friends, who came from Prague and were both involved in the Thun reform in education, made their common career at the University of Vienna for more than thirty years, and when one remembers that Hanslick's masterpiece is dedicated to Zimmermann and that Zimmermann conceives in his *Studien und Kritiken* a whole section entitled the "Musical Laokoon" which meditates on *Vom Musikalisch-Schönen*. The term is thus recurrent in both corpuses and our conviction is that this collaboration between "formalists" was a two-way street: Hanslick modulates his key text, as is well known, but Zimmermann also finds his inspiration in it. It should be noted that this term "phantasy" is also that of his opponents, beginning with Fr. Th. Vischer, to whom he is also opposed by a polemic of more than thirty years.[36] This is how Zimmermann re-appropriates the concept, in his account of the first edition of *Vom Musikalisch-Schönen*:

> It is of little importance that Hanslick, like Vischer, calls the act of contemplation of the objectively beautiful *intuition*, the faculty of this contemplation *phantasy*, since he means nothing but the purely intellectual conception of these conditions on the part of the beholder,

---

34 Wiesing 2008.

35 Landerer 2018, but also Landerer 2004 and 2009.

36 Titus 2016, p. 107f. This revival of Vischer's vocabulary confirms that the aesthetics of form and the aesthetics of content are constantly defining themselves in relation to each other, or even determining each other to the point where they cannot be understood without each other. Cf. the end of my monograph on Zimmermann (Maigné 2017), which analyses the polemics Zimmermann/Vischer.

> through which the disinterested aesthetic judgment of pleasure or displeasure is brought about."[37]

Against a Hegelian meaning of this term, Zimmermann formalizes it, to bring it closer to Herbart's definitions cited above. He makes it an aesthetic and not a metaphysical attitude, a place for the objectification of what is properly aesthetic, between understanding and feeling. Zimmermann also reactivates a concept of imagination stemming from antiquity: it is a question of insisting on the creative and at the same time not subjective capacity of the image, of insisting on the independence of the form with regard to the content represented.

Hanslick defines phantasy as follows: "the agency (*Organ*) by which the beautiful is received is not feeling (*Gefühl*) but rather imagination (*Phantasie*) understood as the activity of pure contemplation" and

> a musical piece emerges from the imagination of the artist for the imagination of the listener. [The activity of] imagination, confronted with the beautiful, is of course not merely a viewing, but rather a viewing with understanding, that is, mental representation and judgment. The latter occurs of course with such speed that the individual processes do not eve n rise to consciousness [...] the word "contemplation", long since extended from the visual to all sensory phenomena, accords fittingly, moreover, with the act of attentive listening, which in fact consists in a successive observing of tone configurations. Imagination is here by no means an isolated domain. Just as it drew its vitality from sensory perceptions, it in turn rapidly transmits its radical waves to the activity of understanding and feeling.[38]

This definition of the beautiful as accompanied by reason but conceived in a fluidity such that we are not aware of its moments, refers to the definition of the beautiful in Bolzano in § 14 of *The Concept of Beauty*, where we see how this Bolzanian thesis was able to combine with Herbartian psychology.[39] At the same time, Hanslick recognizes in Herbart the one who guided him on the path of a formal aesthetics, against an aesthetics of feeling.[40] Phantasy ensures two things in the text: the self-referentiality of the music and the dynamism of its own forms. Thus:

---

37 Zimmermann 1870b, p. 241: "Es thut nichts zur Sache, dass Hanslick den Act der Betrachtung des objectiv Schönen mit Vischer *Anschauung*, das Vermögen desselben *Phantasie* nennt, da er doch nichts als die rein intellectuelle Auffassung dieser Verhältnisse von Seiten des Beschauers meint, durch welche das interesselose ästhetische Urtheil des Beifalls oder Missfallens herbeigeführt wird."

38 Hanslick 2018, chapter 1, p. 13–15.

39 Blaukopf 1996; Maigné 2017a and 2017b.

40 Hanslick 2018, chapter 1, p. 29.

> The ideas that the composer represents are above all and foremost purely musical ones. A particular melody appears in his imagination (*Phantasie*). It should be anything other than itself.[41]

It makes it possible to understand the musical as dynamic,[42] and this dynamic is not a mechanical juxtaposition of elements but comes from "the free creativity" (*Schaffen*) of phantasy.[43] Hanslick sums it all up in a very famous definition: "The content of music is sonically moved forms."[44]

"Aesthetics as a pure science of form is a morphology of the beautiful."[45] General aesthetics is the science of what pleases or displeases, not in the sense of content but in terms of forms, namely relations that are created. Morphology describes the fundamental aesthetic forms whose law is neither that of the world nor that of the subject. Zimmermann insists on the strict distinction between the fact that assessment is carried out *in* the subject and the fact that it is an assessment *by* the subject.[46] The aesthetic imagination is not without rules, without its rules being those of knowledge:

> If according to the laws of pure reason the lawfulness of the imagination was regarded as the origin of all the world of appearance, then the lawfulness of the same without law, the free harmony of the soul-forces was regarded by subjective idealism as the original source of beauty.[47]

> It is the enjoyment of the harmony of the form-giving (concept-forming) and matter-giving (sensual) *soul-power* that distinguishes the aesthetic from the common *imagination*, which creates contentless forms (according to the form-giving power as *thinking*) and formless content (according to the matter-giving power as *intuition*).[48]

---

41 Hanslick 2018, chapter 2, p. 12.

42 Hanslick 2018, chapter 2, p. 15: "What can music portray of feelings, then, if not their content? Only their dynamic properties. It is capable of simulating the motion of a physical process according to the aspects fast, slow, strong, weak, ascending, descending. However, motion is but one trait, one aspect of feeling, not feeling itself."

43 Hanslick 2018, chapter 3, p. 18.

44 Hanslick 2018, chapter 3, p. 4.

45 Zimmermann 1865, § 73.

46 Zimmermann 1865, § 49.

47 Zimmermann 1870a, p. 235: "Galt die Gesetzmässigkeit der Einbildungskraft nach den Gesetzen der reinen Vernunft als Ursprung aller Erscheinungswelt, so galt die Gesetzmässigkeit derselben ohne Gesetz, die freie Eintracht der Seelenkräfte dem subjectiven Idealismus als Urquell der Schönheit."

48 Zimmermann 1870a, p. 234: "Der Genuss der Harmonie der formgebenden (Begriffe bildenden) und stoffgebenden (sinnlichen) *Seelenkraft* ist es, wodurch die ästhetische von der gemeinen

Zimmermann will both maintain indifference to the existence of the object and give an objective rule of taste, which Kant cannot do on the basis of the reflective judgment:

> This setting-free of the imagination is done either at the instigation of an external object, or without it. Kant's assertion that the ultimate purpose of art is to set it free seems to point to the former; for how, for example, can plastic art do this other than by presenting certain external objects? To the latter seems to point the opposite assertion, that we erroneously transfer the source of the natural feeling of pleasure to the external object as the cause of it; for this can mean no more than that the object is quite indifferent to the origin of the feeling of pleasure.[49]

But this objectivity needs the strict distinction between aesthetics and psychology:

> The image of phantasy belongs to aesthetics, not the *fact*. To prove the fact as *given* is up to *psychology*. Aesthetics is therefore not concerned with the question of whether phantasy produces or merely reproduces according to the matter. Aesthetics indeed does not concern the matter of phantasy, the content of conceiving, its *quid*, but only the *form* of the matter, through which it becomes a conceiving full with phantasy, it does approach only its *how*.[50]

It is important to note that the double heritage of Bolzano and Herbart is played out in this quotation. Bolzano exhibits, precisely on the model of music, works of art that exist first in thought before being performed, and which can also subsist as mere thoughts. Bolzano, whose concept of the work of art of mere representation (*Künste der blossen Vorstellung*) is also reinterpreted by Zimmermann. Bolzano explains: "Every work of art must be something real but it does not have to be an

---

Imagination sich unterscheidet, welche entweder überwiegend nach der formgebenden als *Denken*, oder nach der stoffgebenden als *Anschauen*, inhaltsleere Formen und formlosen Inhalt schafft."

49 Zimmermann 1870a, p. 228f.: "Entweder dies in Freiheit-Setzen der Einbildungskraft erfolgt auf Veranlassung eines äusseren Objectes, oder ohne dasselbe. Auf das erstere scheint Kant's Behauptung zu deuten, dass es der Endzweck der Kunst sei, dieselbe in Freiheit zu setzen; denn wie vermöchte dies z. B. die plastische Kunst anders als durch Vorführung bestimmter äusserer Gegenstände? Auf das zweite dagegen die entgegengesetzte, dass wir die Quelle des natürlichen Lustgefühls *fälschlicherweise* auf das äussere Object als Grund desselben übertragen; denn dies kann nur soviel heissen, als der Gegenstand sei ganz gleichgiltig für die Entstehung des Lustgefühls."

50 Zimmermann 1865, § 367: "Das Bild der Phantasie gehört der Aesthetik an, nicht die *Thatsache*. Diese, als *gegebene* nachzuweisen, ist Sache der *Psychologie*. Die Aesthetik bekümmert sich daher weiter auch nicht um die Streitfrage, ob die Phantasie *producire* oder lediglich *reproducire* dem *Stoffe* nach, weil sie eben der *Stoff* der Phantasie, der Inhalt des Vorstellens, dessen *Was* nicht, sondern nur die *Form* des Stoffs, wodurch er phantasievolles Vorstellen wird, sein *Wie* angeht."

object that exists in external reality, that is to say an object that can be perceived by the external senses."[51] This phantasy does not respond to natural laws, but to aesthetic norms, which does not prevent its also being a fact of psychology. The anti-psychologism of aesthetic judgment is based on a psychology that has become a science, that accounts for the dynamics of representations. Is Zimmermann surreptitiously reintroducing a psychology of the faculties? It doesn't seem so. Phantasy is not a power, but an aesthetic modality of knowing: "The one who has phantasy does not merely represent, as the one without phantasy does too, but he represents differently, he represents in a way that is itself aesthetic."[52]

Phantasy describes the work but also the spirit of the one who creates:

> The common conceiving proceeds according to natural laws, the conceiving full of phantasy conforms in addition to aesthetic norms. [...] It (the aesthetic head) itself is a work of art, an aesthetically formed spirit, and thus fundamentally different from the not-art works, the aesthetically unformed spirits [...] But his representations are different from ours; they have passed through the purifying form-fire of the phantasy, are alive, important, regulated, objective, animated, they are works of art in miniature.[53]

This dimension matters because Zimmermann associates here the possible objectivity of judgment with the impersonality of the creative process itself: all "aesthetic head" is regulated, and as such speaks to all. It is not a question here of the genius who would receive his rules from the mysteries of nature. Zimmermann, as a Bolzanian and not as a Kantian, acknowledges that even the genius learns the rules, a position also promoted by Herbart.[54] It is rather a question of thinking about a qualitative and quantitative intensification of the form that the artist possesses and that distinguishes him. It is also a question of reflecting on the diffusion of phantasy: "This is how the phantasy transforms the common conceiving by destroying everything common in it. [...] What touches it becomes gold through

51 Bolzano 1849, § 11.

52 Zimmermann 1865, § 341: "Der Phantasievolle stellt nicht bloss vor, wie der Phantasielose auch, sondern er stellt anders, er stellt in einer Weise vor, die selbst schon ästhetisch ist."

53 Zimmermann 1865, § 368: "Das gemeine Vorstellen geht nach Naturgesetzten vor sich, das phantasievolle entspricht obendrein noch ästhetischen Normen. [...] Er (der ästhetische Kopf) selbst ist ein Kunstwerk, ästhetischer geformter Geist, und dadurch grundverschieden von den Nicht-Kunstwerken, den nicht ästhetisch geformten Geistern [...] Aber seine Vorstellungen sind andere, als die unsern; sie sind durch das läuternde Formfeuer der Phantasie hindurchgegangen, sind lebendig, bedeutend, geregelt, objectiv, beseelt, sie sind Kunstwerke im Kleinen."

54 Herbart 1822b, SW, V, p. 91–122: Herbart asserts, p. 99, that a genius would know his rule if psychology were scientific.

the way in which it touches it."[55] Phantasy tears from the common, it shapes it. It is all the more effective precisely because it is not the application of a logical principle: it cannot deduce its rules from understanding (§ 369). Bolzano's influence seems obvious to us here: the beauty is what could be explained, but is not, and it is beautiful precisely because it is obscure but potentially clarifiable.[56]

Zimmermann seeks to energize his morphology of the beautiful, and tries to do so not through feeling, extra-aesthetic (since extra-musical), but through a dynamism immanent to the forms themselves. Zimmermann re-uses this vocabulary of vitality in the horizontality of an aesthetic without transcendence. The work of phantasy is aesthetic, because it introduces a principle of regularity into chaos, it shapes: "Only phantasy recasts the given forms without regularity and without symmetry into formed (*gebildet*) and symmetrically regulated forms."[57] Phantasy produces harmony and regularity, without the need to incarnate in an existing object. Therefore, it creates a bond between people, regardless of the reality that surrounds them.

3 An Aesthetic Culture of Collective Images?

The formal Herbartian aesthetic and Zimmermann's in particular wants to reinvent a *Bildung*. The science of form is not only an epistemology of art, aesthetics is here an applied ethics. Zimmermann inherits this Herbartian position which he makes his own: ethical judgment is a special case of aesthetic judgment. In Herbartism there is a deep link between the refusal of transcendental freedom, the importance of pedagogy and the reevaluation of aesthetics:

> Philosophical systems in which one accepts either fatalism or transcendental freedom exclude themselves from pedagogy. For they cannot accept without inconsequence the concept of educability (*Bildsamkeit*), which denotes a passage from indeterminacy to firmness.[58]

Since form is at once what creates the link with others, beyond the vagaries of each individual's singularity, and what emancipates one from oneself, one's environment, one's heritage, in a search with inevitably political accents in a

---

55 Zimmermann 1865, § 370: "So bildet die Phantasie das gemeine Vorstellen um, indem sie alles Gemeine daran vernichtet. [...] Was sie berührt, wird zu Gold durch die Art, wie sie es berührt."

56 Bolzano 1843, § 10–14.

57 Zimmermann 1865, § 371: "nur die Phantasie schmilzt die regellos und unsymmetrisch gegebenen Formen zu gebildeten und symmetrisch geregelten um".

58 Herbart 1831, SW IX p. 69.

multicultural empire. The hypothesis of our work is that the logic of the image in Zimmermann fulfills this function and that phantasy is a place of education of oneself and others in a social and historical universe that never ceases to think of its stability.

> The spirit alongside spirits in a communal sense-world, which enables the appearance of the one to the other, is the *social* spirit. The Aesthetics refuses to prove its reality; it prepares itself to design the form of the social spirit insofar as it thereby attains pleasantness, i. e. becomes *beautiful* as a social spirit. Aesthetics asks how this spirit should appear; *that* it should appear, and how it *achieves* that it should appear to the other by means of the sensory world, aesthetics knows from the science of the *spirit* as being, from *psychology*, and from the sensory world as being, from *natural science*. It is in the nature of the spirit that it embodies itself where its matter is given to it; it is in the nature of sensual matter to embody itself, i. e. how far it can represent itself in it.[59]

He is leaning on Herbart again. The long introduction of volume 2 of Herbart's *Psychologie als Wissenschaft* devotes over fifty pages to a state-psychology (*Staatspsychologie*). The latter is also available in state-statics (*Statik des Staates*) and state-mechanics (*Mechanik des Staates*), and must respond to an art of the state (*Staatskunst*), which is in line with Herbart's extreme extension of the term "art". The collective dimension is not only a possible extension of individual psychology, it is intrinsic to it. The dynamic and monadological character of the dynamism of the representations explains why Herbart does not envisage a real break between the individual and the collective scale. The concept of the state is here non-Hegelian: not a realization of the absolute, but on the contrary, a fragile balance of moving representations. There is a threshold of consciousness that is the threshold of social influence (*Schwelle des gesellschaftlichen Einflusses*), where a mass of low-powered individuals remains powerless (*unwirksam*) in their entirety.

In addition to this, attention and insight already mentioned: earning the form is learning an otherness to oneself, it is emancipating oneself from the immediacy of oneself to oneself. The group dynamic takes this up from the individual, so that cooperation and competition are worthwhile here as well. It is obvious that the

---

59 Zimmermann 1865, § 680: "Der Geist neben Geistern in einer gemeinschaftlichen Sinneswelt, welche die Erscheinung des Einen für den Anderen ermöglicht, ist der *sociale* Geist, dessen Realität zu erweisen die Aesthetik von sich ablehnt, dessen Gestalt, insofern er dadurch Wohlgefälligkeit erlangt, d. i. als socialer Geist *schön* wird, zu entwerfen, sie sich anschickt. Wie er erscheinen solle, frägt sie; *dass* er erscheine, und wie er es *bewerkstellige*, dass er mittels der Sinneswelt dem Andern erscheine, entnimmt sie der Wissenschaft vom *Geist* als seiendem, der *Psychologie*, und von der Sinneswelt als seiender, der *Naturwissenschaft*. In der Natur des Geistes liegt es, dass er sich verkörpert, wo ihm sein Stoff gegeben ist; in der Natur des sinnlichen Stoffs liegt es, wie weit er sich verkörpern d. h. in demselben sich abbilden kann."

psychology of interactions is also a psychology of possible failures: the psychological basis of aesthetics is not a philosophy of history and its linear or dialectical progress, it is a dynamic of depths. Jäger has explained very well that *Tiefenpsychologie* takes the place of a philosophy of history in Herbartism: the stabilization of a representation and the adherence to this representation are acquired, unstable, always subject to a dynamic that is both conscious and unconscious.[60]

There are aesthetic features of living together, those of any work. If indeed, "The appearance of the spirit for itself, because itself only representation, is *ideal*; the appearance of the spirit for the other(s), expressed in sensual matter, *real*."[61] In the same way that the subject in Herbart's work becomes a center of our representations,[62] each socially understood individual is articulated here to others, is traversed from one side to the other by the collective. Zimmermann insists on the fact that aesthetics, not knowing and wanting to know anything about being, has no interest in knowing what are the actual national and historical links between the men of this brotherhood. His formalism is intended to be supranational and supra-historical, faithful in this respect to Bolzano. He is convinced that it is possible to create an aesthetic community, that is to say, according to him, of the duty to be and not of being, of choice and not of being. Zimmermann here engages a social imaginary, a work of *Socialphantasie* (§ 733). Certainly, it is not about the real, but it is perhaps about the *sine qua non* condition of the transformation of the real: that of the imaginary as a field of possibilities before being. There is a work of the social imaginary in the work of art. The aesthetic society is thus a society of mirroring oneself in others, a process of recognition and progressive equality.[63] Arens rightly insists on the way Zimmermann re-evaluates Herder's theses in his *Geschichte der Aesthetik*, elevating Kalligone to the rank of an aesthetic work capable of countering Kantism, concerned with capturing an experi-

---

60 Jäger 1982, p. 199: "Seine Lehre kennt keine Natur- und Geschichtsphilosophie, die die Selbstrealisation Gottes oder die Offenbarung der Idee im Werden der Dingen verfolgt. An die Stelle der geschichtlichen Behandlung tritt in seiner Schule die Psychologie als Grundwissenschaft. Die Bedeutung dieser Sache für die österreichische Geistesgeschichte wird m. E. noch nicht voll erkannt. Die eröffnet auch erst das Verständnis der Formalästhetik."

61 Zimmermann 1865, § 683: "Die Erscheinung des Geistes für sich selbst, weil selbst nur Vorstellung, ist *ideal*; die Erscheinung des Geistes für den oder die Andern, im sinnlichen Stoff ausgeprägt, *real*."

62 Herbart 1829, SW VIII, § 325, p. 232: "It is enough for the eidolology to know that the ego is and can be nothing other than a centre point of changing representations." ("Der Eidolologie genügt es zu wissen, dass das Ich nichts Anderes ist und sein kann, als ein Mittelpunct wechselnder Vorstellungen.").

63 Zimmermann 1865, § 728.

ence that is not made of pure representations (*Nachbilder*) but of images (*Bilder*), therefore of mixed experience, including judgment and the body, in order to aim at the well-being of humanity.[64]

Art becomes for him a place of "analytical experimentation" and a "terrain for the exercise of ideal sociality", a perspective deployed by Herbartism as a whole.[65] Zimmermann does not only question the internal structure of the work and its realization in the representation of the recipient, but considers the work as an interpersonal entity of relationship and communication.[66] Is the term semiotics appropriate? It reintroduces two problematic things with regard to Zimmermann's text: it introduces a content and an explication of the content. Zimmermann precisely proposes an aesthetics where the implicit gives form without ever giving reasons.

Zimmermann extends aesthetics beyond the fine arts. The term *Kunst* is thus extended far beyond the masterpiece to encompass the whole of what organizes life in society. Psychological perception and a theory of language complete the logical analysis of beauty:

> Real art, as the language of the beautiful spirits among each other, mediates the phantasy of the one with that of the other, and raises it to the common, *social phantasy*, like *the science presented visibly or audibly* (literature and school), the language of the spirits directed at the recognition among each other of what is right and valid, mediates the conviction of one with that of the other, and raises it to the communal, the social conviction, to the *social knowledge*. Since both, insofar as they represent thoughts, depend on *artificial* signs, they are also exposed to constant dangers of misunderstanding in their mediation among the spirits. The real, visible and audible work of art, as well as the knowledge that has been communicated with it, can, as a result of the defective sign, be interpreted differently than they were intended, and the spirits appearing through it as *different* to the others than they intended, or they can also do so *because* they wanted to, and deliberately make use of the crack that the artificial sign covers, but does not remove.[67]

---

64 Arens 2016, p. 135, 140, 145.

65 Schneider 2009, p. 95.

66 Schneider 2009, p. 93f.

67 Zimmermann 1865, § 760: "Die reale Kunst als die Sprache der schönen Geister unter einander vermittelt die Phantasie des Einen mit der des oder der Andern und erhebt sie zur gemeinschaftlichen, zur *socialen Phantasie*, wie die *sicht-* oder *hörbar dargestellte Wissenschaft* (Literatur und Schule), die Sprache der auf das Erkennen des Richtigen und Giltigen gerichteten Geister unter einander, die Überzeugung des einen mit der des oder der Andern vermittelt und sie zur gemeinschaftlichen, zur socialen Überzeugung, zum socialen *Wissen* erhebt. Da beide, insofern sie Gedanken darstellen, auf *künstliche* Zeichen angewiesen sind, so sind sie in ihrer Vermittlung unter den Geistern auch beständigen Gefahren des Missverständnisses ausgesetzt; das reale

What this quotation, divided into two distinct moments, shows is that it is precisely language that can blur the reception of the work, and is not necessarily explicit about it. The evidence of the aesthetic judgment is without words. It is precisely capable of being posed without a theoretical or theorized universal. It is precisely not based on a concept of harmony or unity (metaphysics). Art becomes a medium because the constraint of judgment exists, the aesthetic judgment having an objectivity, imposing itself according to it. But this objectivity is not logical: whereas logical judgment can have any predicate, aesthetic judgment has only one, the feeling of pleasure or displeasure.

Herbart has already been able to underline how much apperception is part of understanding monuments, in other words history. Explicitly, Herbart specifies that it is not enough to perceive them but to apperceive them, that is to say to appropriate them:

> The impression of old buildings, considered as monuments, shows very clearly what amounts to apperception in works of art; an apperception which is far different from mere *perception*, included the artistic impressions based on this later one (*Psychology* II § 125 et seq.). With what eye the historian looks at an old coin! A historical *appropriation* (and the apperception is nothing else) gives it its value.[68]

Aesthetics is a work of the self on the self by purifying the self in the name of a norm that imposes itself by its form. This emancipation thanks to the form must not be a work of artifice (*Künstelei*), but rather a stylization of the subject and the others[69]. What is surprising, without breaking with Herbart at all, is the way in which the whole everyday life of a culture is inserted into these aesthetic forms: Zimmermann will successively analyze types of school (*Volkschule, Gelehrtenschule, Universität*), the relations between teachers and servants, between men and women, between the believer and his church, etc. The whole of political life is grasped, as a public conscience based on constituted bodies, objects of the

---

sicht- und hörbar gewordene Kunstwerk wie das mitgetheilte Wissen können in Folge des mangelhaften Zeichens anders gedeutet werden als sie gemeint waren, die dadurch erscheinenden Geister als den Andern *anders* erscheinen als sie es beabsichtigten, oder sie können dies auch, *weil* sie es wollten, und sich des Risses, welchen das künstliche Zeichen wohl verdeckt, aber nicht aufhebt, dazu mit Vorsatz bedienten."

68 Herbart 1831, SW IX, § 70, p. 107: „Der Eindruck alter Bauwerke, die als Denkmäler betrachtet werden, zeigt es recht deutlich, wieviel bei Kunstwerken auf die *Apperception* ankomme, die von der blossen *Perception*, sammt den auf ihr allein beruhenden Kunst-Eindrücken, weit verschieden ist (*Psychologie* II § 125 u.s.w.). Mit welchen Augen sieht der Historiker eine alte Münze! Eine historische *Aneignung* (und nichts anderes heisst *Apperception*) gibt ihr den Wert."

69 Zimmermann 1882, § 399.

dynamic psychology of representations. This life is understood as a public conscience based on constituted bodies, objects of the dynamic psychology of representations: school, church, state, army, political party, but also the press, the caricature etc.[70] Very clearly, aesthetic feelings socialize, create empathy, adhesion, invent a feeling of humanity (*Humanitätsgefühl*)[71] that should be promoted against anti-social impulses. Zimmermann therefore proposes to extend the aesthetic grasp of the human world in all its manifestations, opening up here towards sociological or anthropological analyses that would be as much an awareness of what is given in a latent way in daily life. The formal aesthetics that he develops has such an extension that it becomes a reflection on the whole of what makes up individual and collective life. In this sense, it is very much part of the "system of culture" mentioned in the introduction: it is both description and self-reflexivity, Herbartism works culture as it works concepts.

References

Arens, Katherine (2016), "Rereading Herder as an Heritor of Idealism: Robert von Zimmermann's Aesthetics", in: *Herder Jahrbuch* XIII, Heidelberg: Synchrone, p. 129–146.

Blaukopf, Kurt (1996), *Die Ästhetik Bernard Bolzanos*, St. Augustin: Akademia.

Bolzano, Bernard (1843), *Über den Begriff des Schönen*, reprint in: *Gesamtausgabe*, Reihe 1, Bd. 18: *Mathematisch-Physikalische und Philosophische Schriften 1842–1843*, G. Gabriel/M. Gatzenmeier/F. Kambartel (eds.), Stuttgart/Bad Cannstatt: Fromann-Holzboog 1989, p. 87–238.

Bolzano, Bernard (1849), *Über die Einteilung der schönen Künste*, reprint in: Bolzano, *Untersuchungen zur Grundlegung der Ästhetik*, D. Gerhardus (ed.), Frankfurt: Athenäum Verlag 1972, p. 119–273.

Bolzano, Bernard (2017), *Écrits esthétiques*, C. Maigné/J. Sebestik/N. Rialland (eds.), Paris: Vrin.

Hanslick, Eduard (2018), *On the Musically Beautiful*, new translation by L. Rothfarb and Ch. Landerer, Oxford: Oxford University Press.

Henckmann, Wolfhardt (1993), "Einleitung", in: J. F. Herbart, *Lehrbuch zur Einleitung in die Philosophie*, W. Henckman (ed.), Hamburg: Meiner, p. VII-LIX.

Henckmann, Wolfhardt (2001), "Über die Grundzüge von Herbarts Ästhetik", in: A. Höschen/L. Schneider (eds.), *Herbarts Kultursystem. Perspektiven der Transdisziplinarität im 19. Jahrhundert*, Würzburg: Königshausen & Neumann, p. 231–258.

Herbart, Johann Friedrich (SW), *Johann Friedrich Herbart's Sämtliche Werke in chronologischer Reihenfolge*, K. Kehrbach/O. Flügel (eds.), 19 vol., Langensalza, 1887–1912. Reprint Aalen: Sciencia Verlag 1964 (SW, following by volume).

70 Zimmermann 1882, § 412.
71 Zimmermann 1882, § 413.

Herbart, Johann Friedrich (1802), *Pestalozzi's Idee eines ABC der Anschauung*, SW I, p. 151–274.
Herbart, Johann Friedrich (1808), *Allgemeine praktische Philosophie*, SW II.
Herbart, Johann Friedrich (1822a), *De attentionis mensura causisque primariis*, SW V, p. 41–89.
Herbart, Johann Friedrich (1822b), *Über die Möglichkeit und Notwendigkeit, Mathematik auf Psychologie anzuwenden*, SW V, p. 91–122.
Herbart, Johann Friedrich (1825), *Psychologie als Wissenschaft* II, SW VI.
Herbart, Johann Friedrich (1829), *Allgemeine Metaphysik* II, SW VIII.
Herbart, Johann Friedrich (1831), *Kurze Encyclopädie*, SW IX.
Herbart, Johann Friedrich (1836), *Analytische Beleuchtung des Naturrechts und der Moral*, SW X.
Herbart, Johann Friedrich (1837), *Lehrbuch zur Einleitung in die Philosophie*, W. Henckman (ed.), Hamburg: Meiner 1993.
Jäger, Georg (1982), "Die Herbartianische Ästhetik. Ein österreichischer Weg in die Moderne", in: H. Zeman (ed.), *Die österreichische Literatur, ihr Profil im 19. Jahrhundert (1830–1880)*, Graz: Akademische Druck- und Verlagsanstalt, p. 195–219.
Landerer, Christoph (2004), Eduard Hanslick und Bernard Bolzano. Ästhetisches Denken in Österreich in der Mitte des 19. Jahrhunderts, St. Augustin: Academia.
Landerer, Christoph (2009), "Hanslick et le formalisme en Autriche", in: Maigné/Trautmann-Waller (2009), p. 73–84.
Landerer, Christoph (2018), "Introductory Essays", in: Hanslick 2018, p. 1–54.
Maigné, Carole (ed.) (2012), *Formalisme esthétique. Prague et Vienne au XIXᵉ siècle*, Paris: Vrin.
Maigné, Carole (2017a), *Une science autrichienne de la forme. Robert Zimmermann (1824–1898)*, Paris: Vrin.
Maigné, Carole (2017b), "L'exigence de la clarté. Les *Ecrits esthétiques* de Bernard Bolzano", in: B. Bolzano, *Écrits esthétiques*, Paris: Vrin, p. 7–46.
Maigné, Carole (2018), "Herbarts ABC der Anschauung und der ästhetische Formalismus", in: J.-F. Goubet/R. Bolle (eds.), *Herbart als Universitätslehrer*, Jena: Paideia, p. 75–88.
Maigné, Carole/Trautmann-Waller, Céline (eds.) (2009), *Formalismes esthétiques et héritage herbartien. Vienne, Prague, Moscou*, Hildesheim/Zürich/New York: Olms.
Schneider, Lothar (2009), "Quelques figures de l'esthétique herbartienne", in: Maigné/Trautmann-Waller (2009), p. 85–100.
Stout, George F. (1891), "Apperception and the Movement of Attention", in: *Mind* 16, no. 61, p. 23–53.
Titus, Barbara (2016), *Recognizing Music as an Art Form: Friedrich Th. Vischer and German Music Criticism 1848–1887*, Leuven: Leuven University Press.
Wiesing, Lambert (2001), "Formale Ästhetik nach Herbart und Zimmermann", in A. Höschen/L. Schneider (eds.), *Herbarts Kultursystem, Perspektiven der Transdisziplinarität im 19. Jahrhundert*, Würzburg: Königshausen & Neumann, p. 283–296.
Wiesing, Lambert (2008), *Die Sichtbarkeit des Bildes. Geschichte und Perspektiven der formalen Aesthetik*, Frankfurt am Main: Campus Verlag.
Zimmermann, Robert (1858), *Aesthetik. Erster, historisch-kritischer Teil: Geschichte der Aesthetik als philosophische Wissenschaft*, Vienna: Braumüller.
Zimmermann, Robert (1861), *Philosophie und Erfahrung. Eine Antritsrede*, Vienna: Braumüller.
Zimmermann, Robert (1862), "Zur Reform der Aesthetik als exakter Wissenschaft", in: *Zeitschrift für exakte Philosophie* 2, p. 309–358.

Zimmermann, Robert (1865), *Aesthetik. Zweiter, systematischer Teil: Allgemeine Aesthetik als Formwissenschaft*, Vienna: Braumüller.
Zimmermann, Robert (1870a), "Zur Reform der Aesthetik als Kunstwissenschaft", in: Zimmermann, *Studien und Kritiken* I. *Zur Aesthetik*. Translated into French by M. Galland-Symkoviak and J.-O. Begot, in: Maigné 2012, p. 223–265.
Zimmermann, Robert (1870b), "Vom Musikalisch-Schönen", in: *Österreichische Blätter für Literatur und Kunst* 11, 1854; in: Zimmermann, *Studien und Kritiken* II. *Zur Aesthetik*, Vienna: Braumüller, p. 239–253.
Zimmermann, Robert (1882), *Anthroposophie im Umriss. Entwurf eines Systems idealer Weltansicht auf realistischer Grundlage*, Vienna: Braumüller.

Index of Names

https://doi.org/10.1515/9783110747324-008